CONSTRUCTION SAFETY AND LOSS CONTROL PROGRAM MANUAL

CONSTRUCTION SAFETY AND LOSS CONTROL PROGRAM MANUAL

ANDREW CIVITELLO, JR.

SP

SHARPE PROFESSIONAL

An imprint of M.E. Sharpe, INC.

This publication is designed to provide accurate information in regard to the subject matter covered. It is sold with the understanding that the publisher is not engaged in rendering legal, accounting, or other professional service. If legal advice or other expert assistance is required, the services of a competent professional person should be sought.
—*From the Declaration of Principles adopted by a Committee of the American Bar Association and a Committee of Publishers and Associations.*

Library of Congress Cataloging-in-Publication Data

Civitello, Andrew, Jr., 1954–
Construction safety and loss control program manual / by Andrew Civitello Jr.
p. cm.
Includes index.
ISBN 0-7656-0181-8 (cloth : alk. paper)
1. Building—Safety measures. 2. Construction industry—Safety regulations.
3. Business Losses. I. Title.
TH443.C58 1997
690′.22′0973 dc21 97-30499
CIP

Printed in the United States of America

The paper used in this publication meets the minimum requirements of the American National Standard for Information Sciences—Permanence of Paper for Printed Library Materials, ANSI Z 39.48-1984.

∞

BM (c) 10 9 8 7 6 5 4 3 2 1

Disclaimer

IMPORTANT—The author and publisher do not possess any special safety certifications, licenses, or other implicit authority. The author has employed the use of such experts to varying degrees in the preparation and review of the materials presented, but is not representing himself to be such a safety expert.

The introductory sections of the *Construction Safety and Loss Control Program Manual* are intended to explain that the manual is a sample of the final product resulting from a company's effort to apply the various rules and regulations affecting the safety and health of all employees. It is to be considered as an example of the manner of presentation, and not as an absolute guide with respect to material content.

Information for the manual has been secured from many sources. The complete *Construction Safety and Loss Control Program Manual* as it is presented may ultimately include either too much or not enough material for your own organization.

Prior to the issue of your actual safety and loss control program manual for use by your own company and employees, the services of competent and legally authorized safety, legal, and human resources professionals in your state should be employed to consult with and review and advise on the preparation of, and confirm all information. And while all of the herein information is believed and thought to be accurate, neither the author nor the publisher assumes any liability either express or implied in connection with the use and application of this information.

Acknowledgments

This *Construction Safety and Loss Control Program Manual* is dedicated to the men and women in our industry who have taken upon themselves the responsibility to educate and train our industry members in safe work practices, and who have dedicated their efforts toward achieving real, significant, and ongoing improvements in the safety and health conditions of our jobsites for all the right reasons.

About the Author

Andrew Civitello, Jr. has managed construction projects of many sizes and types for clients since 1976 that include local, state, and federal government agencies and departments, municipalities, banks, service organizations, and private developers. He is currently Senior Construction Consultant for PinnacleOne, a major international construction consulting firm, and is based in their Clearwater, Florida office.

A graduate of Syracuse University, he has been a project manager for major construction companies and president of a Connecticut general contracting firm oriented to public projects. He is also an independent consultant for the disciplines of planning and scheduling, changes/claims, and project management for clients that include general contractors, subcontractors, attorneys, design professionals, and owners. He is a university instructor for project management subjects, an arbiter for the American Arbitration Association, and an expert witness for scheduling, estimating, and claims.

He is the author of *Complete Contracting* (McGraw-Hill 1997), *Construction Operations Manual of Policies and Procedures* (McGraw-Hill 1994), *Contractor's Guide to Change Orders* (Prentice Hall 1987), *The Builder's and Contractor's Yearbook* (Prentice Hall 1987), *The Construction Manager* (1987 through current year), and co-author of *Construction Scheduling Simplified* (Prentice Hall 1985).

A Special Thanks

A special thanks is given to *Kearney Development Company, Inc.*, in Tampa, Florida. Many sample sections provided in the *Construction Safety and Loss Control Program Manual* were based on materials originally developed by Kearney Development Co., Inc. in their effort to produce a comprehensive program for use by their own employees. Their significant efforts in the areas of safety, employee health, and loss control exemplify the highest standard of care and achievement. Their generous authorization of the use of material incorporated throughout this *Construction Safety and Loss Control Program Manual* is greatly appreciated.

Contents

Introduction

"Safety" has become another one of those buzzwords getting much use in the construction industry at increasing rates. It is a word, however, with real meaning—and equity—that varies from company to company and from individual to individual perhaps more than any other word in the business. It is a word that deserves direct and serious attention, and is certainly one that should not be allowed to become a cliché as the industry moves forward into the next century.

As an industry, providers of construction services have been forced in recent years to wrestle with a large number of serious dilemmas in ways that are strikingly different from those dealing with similar issues in other industries. Until relatively recently, history has confirmed the construction field to be one of the most consistently dangerous for its employees. While there may be a great distance to go with respect to continuing improvements in our business's safety record, it is at least equally true that as an industry, we've achieved more safety gains than any other. These gains are not merely in statistics and real safety achievement, but are in attitude, perspective, and genuine intent as well.

The Motivations to Be Safe

Like so many other categories of responsibility, the issue of overall safety in construction is now prescribed, directed, recommended—and mandated—in ways that genuinely affect each member of every company that provides construction services. We've been told that "worker safety and health protection is a decisive factor in reducing work-related injuries and illnesses," and "Every employer should develop and maintain a safety program . . . that should provide policies, procedures and practices that protect all employees from safety and health hazards." Like it or not, we as constructors are all part of the program, have certain rights within the system, and carry significant responsibilities with us from the office or shop to the jobsite and back each day. For those of us who are doing what we can to take these realities seriously, they mean different things at different times.

Administrators have the responsibility to determine and adequately provide for the requirements of a proper, company-wide, ongoing safety program; managers are responsible for implementing and enforcing compliances by every member of the organization; and employees bear the burden of complying with the clear set of company directives as communicated.

Employees have a right to a safe working environment at least to the minimum standards prescribed by law. Employers also have certain rights with respect to expectations that qualified employees will obey laws and act with the responsibility that their position requires.

Beyond all these things, there is a finite set of reasons why any of us in the construction industry are ultimately motivated to provide for safe working conditions in our jobsite and office environments. The relative importance within the total mix that an individual will place on any one component will vary to wide degrees, but intelligent, responsible people should ultimately conclude that giving safety regular, significant attention is not only the right thing to do, but it is good business as well.

All motivations for giving safety proper attention boil down to three categories: moral, economic, and legal. These motivations are the "components" within the total mix that are given different levels of importance by different individuals. All preaching aside, it almost doesn't matter where any of us are along the total safety-perspective continuum, as long as we manage to reach the correct conclusion for ourselves—that genuine attention to safety is necessary, correct, and even profitable.

Morally, we all know the reasons for safety. We know that providing safe working environments for all employees is not only the right thing to do, but it is intuitively clear that none of us have any right under any circumstances to knowingly and willingly subject another individual—whether an employee, peer, or even a family member—to conditions that may or will adversely affect the health of that individual. If any of us need examples of case law or specific government guidelines to convince us of that general principle, we will likely have other difficulties with the "moral" component.

"Doing the right thing" with respect to jobsite safety, however, can often turn out to be an easy principle in black-and-white cases (should there be a handrail or any kind of fall protection at the edge of that fifth-floor slab?), but can become confusing in those gray areas that permeate the safety issue (am I satisfied with those minimum published government guidelines, or is there something about the situation that still leaves me uncomfortable?).

As it turns out, once the basic legal requirements have been determined and provided for in a specific safety situation, there is a simple test that can routinely be applied to every questionable condition that will help to leave a concerned person with clear direction—and a clear conscience. The test can be considered at two levels. First; consider the condition from the point of view of a safety and health-conscious individual, and determine if you would be satisfied with placing *yourself* under the total set of working conditions that you are about to place another individual under. Even if for some reason you are overly casual about the level of peril that you're willing to tolerate personally, consider the second level; simply ask yourself if you would be satisfied to allow your own child or spouse (of adequate age, training, and experience) to work in the same condition. While not specifically satisfying (or violating) any legal requirement that I am aware of, this simple test has always helped me to personally instantly transform those shades of gray clearly to black or white. In other words, if the conditions are not good enough for my kids, they're not good enough for my employees or peers.

On the economic level of consideration, safety is a set of numbers related to an individual's compensation, and to company profit and loss. Studies and statistics can and have cataloged the direct and indirect economic effects on individuals and organizations. The direct effect on an employee begins with loss of wages. Even if compensated by insurance, the reduced physical capacity related to injury or illness affects every aspect of the individual's life. The indirect effects can multiply exponentially.

For the company, economic losses related to a jobsite accident divide still further among added cost related to interruption, rework, and increased costs. A serious accident on a jobsite stops the project—either partially or entirely—until the condition is dealt with on its immediate (emergency), intermediate (secure area), and long-term (corrective measure) phases. The work is interrupted, and wages continue for every employee involved with each phase of the accident. Jobsite supervision is removed from production considerations and focused on the emergency. Office administrators—and even company officers—are pulled from their ongoing work and rushed to the emergency condition.

After consideration of the jobsite and company interruption is the realization that certain preventable safety emergencies can result in damage to the work, which must then be corrected. An example might be a trench collapse that damages shoring, formwork, or equipment in the vicinity. Simply stated, all this means is that after the emergency has been dealt with, the construction area must be reworked at the added expense of removal and replacement of any damaged areas.

Legal considerations are intertwined with all others every step of the way. Minimum requirements for company safety programs and specific criteria for jobsite conditions can be clearly spelled out sufficiently to allow for scheduled or surprise inspections that can result in large fines for perceived violations. Once cited for a safety violation, be prepared to add (of course) those legal fees—even if the company is ultimately found to have been in compliance all along.

Finally (but not least) are the added insurance costs to companies that are guaranteed to be assessed due to the worsening of the company's safety record as the result of experiencing a serious accident. These costs are often not routinely dealt with by anyone other than senior company management and certain administrators, and can have their significance diminished in the eyes of managers and supervisors unless the cost increases are brought to their attention with the full impact that they deserve. Such increases can run into the tens of thousands of dollars, and can continue literally for years until an improved company safety record is reestablished.

When all the labor, equipment, administrative, and overhead costs to a company are added up for the entire scenario to deal with even a single incident of a serious Jobsite safety emergency, it is not difficult to see that the costs that would have been saved through prevention of the incident would likely surpass the cost of implementing an entire effective company-wide safety program. If that implemented program then goes on to prevent a second serious emergency, it becomes clear that the idea of attention to safety moves decisively into the category of a direct major contributor to the company's bottom line. It is not smoke and mirror accounting, but profit/loss reality. It is genuinely profitable to be safe.

It is all these considerations that are alluded to in the phrase "safety is good business." Safety does turn out to be "good business" for all the right reasons.

"What," but Not "How"

Now that we are all hopefully convinced that proper, complete attention to a comprehensive and effective ongoing company safety program is urgently needed, we face the situation placed upon us to varying degrees by our government, insurance, and legal institutions. That situation is one in which these entities often seem to be satisfied with mandating generalities—that the company must have an "adequate safety program"—the "what"—but then stop short of providing instruction regarding the *specific* safety program that would satisfy them—the "how."

If anyone has ever had their safety program criticized by one of these entities, you might have been surprised with an exchange similar to the following:

THEY: "Your safety program is inadequate." *(Translation: "We intend to hold you liable for every conceivable problem resulting . . .")*

YOU: "Would you please provide me with a specific example (a *sample*) of a safety program for a company like mine that you feel is an 'adequate' (or even 'superior') program?" *(Translation: "I will be happy to do exactly what the industry feels is a superior job with respect to safety. Since you (the insurance/legal institution) know what you don't like, certainly you must know what you do like. Just tell me what it is, and I'll do it.")*

THEY: "Oh, we can't do that." *(Translation: "We are not about to assume the liability for a specific set of instructions; we intend to keep that liability squarely with you.")* "We just know that we don't exactly like what you have." *(Translation: "Even though you may have the absolute best program in the world, we can't admit that to you; if we do, we may let you off a hook or two . . .")*

And so we finally get to the purpose of this book.

While there are many compilations of government requirements and safety standards that catalog various procedures to be implemented in various situations, those criteria actually form only certain components of an overall company safety program. The purpose of this book is to begin the effort to distill the libraries of different (sometimes differing) sets of "instructions," and boil them down to an actual sample safety program that can finally be identified and implemented.

The overall effort is to work with the objective of determining a standard of performance that range from "acceptable" to superior, and to then attempt to develop a comprehensive approach to a company safety program that meets or exceeds those expectations.

This *Construction Safety and Loss Control Program Manual* is a guide and a ready-to-use example. It will hopefully serve as the starting point for the preparation of the specific safety program that will ultimately represent a superior effort for the needs of your particular company. It identifies the components that make up such a superior program for a construction company, and addresses them in ways that meet the needs of the industry as described in this introduction.

What This Manual Will Do For You

The *Construction Safety and Loss Control Program Manual* will provide you with:

- A ready-to-use model of an actual safety and loss control program manual that can be implemented by your company.
- A catalog of the critical safety and loss control issues facing contractors and construction managers today.
- A sample subcontractor safety program that should be included as a minimum requirement in all of your subcontracts.
- Guidance with respect to the industry approach to the safety and loss control issue, with help on how to develop your own approach that will meet or exceed industry and legal expectations of your operation.
- Suggestions that will help to create a corporate culture of safety.
- Tips on how to make all company executives, administrators, managers and supervisors, and employees aware of the critical company need for safe operations.
- Guidance with respect to the determination of program information distribution. Discussion of which employees should be issued and be held responsible for respective portions of safety and loss control information.
- Discussion on how to keep safety and loss control responsibilities and liabilities where they belong; how to avoid unnecessarily assuming additional significant liabilities both professionally and personally.

How to Use This Manual

Sample Section Format

Each sample section of the *Construction Safety and Loss Control Program Manual* is preceded with a “Section Use” preface. It describes:

- The purpose and objectives of including the sample section in the company’s actual safety and loss control program manual;
- The section’s content description, and how it meets the sample section objectives;
- Relevant considerations; how the section may correlate with other sections in the sample program manual;
- When appropriate, references to additional source, related, or supplemental materials; and
- The steps that need to be taken to customize the sample section to meet the specific needs of your company.

The “Section Use” preface describes the basis for the development of the sample section that immediately follows, but is *not* part of the sample section itself. The section then goes onto become a specific example of how the respective topics might appear in your company’s actual safety and loss control program manual.

The “Section Use” portion is the “how-to” and “why” of the section. The sample section that immediately follows is written in the style of the final program manual itself, as if written by your company as instruction to your company employees. To this end, it is intended that wherever the term “(**COMPANY**)” appears, the name of your own company will be substituted when the draft of the manual section is prepared.

Procedure for Manual Development

The procedure for preparation and development of your own manual begins with a reproduction of the respective sample sections that you may decide will actually serve as the first draft of your actual manual. The material will then be edited in consideration of the issues and concerns that may be identified in the "Section Use" preface. Finally, the completed manual draft should be reviewed by a qualified safety professional, and perhaps by your insurance carrier and attorney as well.

In order to accomplish this, follow these steps:

Step 1. Thoroughly review the Table of Contents, and begin to develop an outline of topics that you may feel should become part of your own program and manual. This material clarifies the complete program intent, content of the materials provided, and key uses.

Step 2. Compare the sample program content as presented for overall construction operations with the specific operations of your own company. A specialty subcontractor (such as an electrical, HVAC, or site contractor), for example, will have needs that are different from those of a general contractor. This consideration may lead you to conclude that certain sample manual sections are adequate for your own operation as they are, or may identify areas that will require a different or more intensive approach for your particular operation.

Step 3. Use the material to draft your own complete *Safety and Loss Control Program Manual. NOTE: The* Quick➔Disk *provided with the sample manual may be used to save valuable time in preparing the first draft material.* Address each section and make your own decisions regarding their relevance to your own operations. Be sure to substitute your own company name in each area with the "**(COMPANY)**" reference, and delete the word "sample" wherever it appears.

Step 4. Be aware that certain decisions may have been made in the preparation of sample section material content that are based in policy and even personal preference.

For example: Deciding whether or not to provide (make available training for all company employees in CPR and basic aid.

Questions:

1. Is this an employer obligation?
2. What, if any, are the liabilities assumed by:
   - the employer?
   - the employee?

in providing aid or care to:

- company employees?
- non-employees on a jobsite?

3. Are there "good Samaritan" or other laws that will protect the company and/or individuals from claims resulting from any efforts to provide emergency care?

4. Is the company or employee liable if an employee is "qualified" to provide emergency care, but does *not* provide the care for any reason?

These types of considerations are highlighted in the text with a ***"NOTE TO MANUAL PREPARER"*** that calls your attention to the respective issue. Consider the material that immediately follows very closely, and make your own decisions regarding the manner in which you intend to treat the respective material in *your* company.

Step 5. Review critical issues with your attorney and insurance carrier in order to confirm the circumstances of your specific situation to the degree necessary to allow an informed management decision in the final determination of company policy with respect to those items.

Step 6. Review your specific program content with your own insurance carrier. Send them your complete draft manual. Request their review and written remarks.

Step 7. Use the remarks from your insurance carrier as your guide. Keep records of your communications in this regard. It will be proof of your own efforts to identify and accommodate the requirements of the industry. This review process will make it that much more difficult for those same insurance carriers to later criticize the adequacy of your written program (note, however, that this won't necessarily address problems relating to a system or communication failure, only program design).

Step 8. Prepare the final manual for reproduction and distribution. Use the *Sample Page Layout* provided at the end of this section to properly identify and organize the material. Consider *Sample Section 1.5 Manual Issue, Revisions, and Use,* and determine who in the organization must be issued either the entire manual, or relevant portions of the ultimate manual configuration allows it to be partitioned. Consider the fact that even though it may be more expensive to reproduce the entire manual for distribution to all employees, such an approach may ultimately turn out to be the cleanest option that will leave no question.

Step 9. Be sure to have each employee execute and return some form of Employee Acknowledgment similar to the sample provided following these steps, on page xliii. Preserve the company's ability to demonstrate its efforts to comply with all regulatory requirements and standards by identifying the precise information that has been issued to specific individuals.

Step 10. Enforce strict procedure to preserve accountability among all company administrators, supervisors, and employees. Be sure that all company participants realize the hard work that has gone into the preparation and dissemination of the program, and that the final product is an extremely important set of organized instructions *that the company will use to measure the performance of its employees*. Be certain to notify everyone that the final material is proprietary in nature, is the exclusive property of the company, and that you unequivocally expect that each employee give the issue the attention and respect that it deserves.

Additional Suggestions

Beyond the issues, ideals, and objectives presented and explored in the effort to create a thoroughly comprehensive company *Safety and Loss Control Program Manual*, the safety effort of the organization should become an ongoing integral component of each employees' consciousness. It should include a learning and development process that never stops. Look for opportunities to keep the knowledge and implementation cycles about safety spiraling upward. Consider these suggestions as part of ongoing safety consciousness:

1. Visit your local office of the Occupational Safety and Health Administration (OSHA). You are certain to find exceptional staff who will be extremely helpful in directing you toward a wealth of information prepared for your specific type of operation. You'll find videos, printed materials, and information on available training programs.
2. Visit your local library. Locate industry directories of individuals and companies certified in various safety subjects, who may be available for consultation regarding the preparation of your custom safety and loss control program, or to provide individual inspection, training, or other services that will support and enhance your safety efforts company-wide.
3. Identify sources of available safety training in basic first aid, CPR, advanced first aid, and so on. Consider the local chapter of the American Red Cross, local fire departments, safety training companies, and certified individuals.
4. Review the Appendix to this *Construction Safety and Loss Control Program Manual.* The Appendix is provided as a supplement to the complete sample *Construction Safety and Loss Control Program Manual* as it has been presented. It is a compilation of selected training requirements and guidelines that have been prepared by the U. S. Department of Labor Occupational Safety and Health Administration specifically to assist companies in the distillation process of converting stated compliance requirements of the OSHA standard into practical operating activities. The U.S. Department of Labor has provided these types of material guidelines for many industries, and in a manner that addresses working conditions throughout many regional environments. This Appendix has been developed in an effort to improve the efficiency of the use of this type of information for the members of the construction industry, and is the result of isolating and consolidating those requirements and guidelines that directly address our own construction industry.

Sample Page Layout

The Sample Page Layout below is an example of a format that should be used in the entire manual. Certainly, you are free to adjust the page form in its appearance and content, but such modifications should be made in consideration of the following:

COMPANY LOGO	Construction Safety and Loss Control Program Manual	Page: 1-1	Appr:
	1. CONTENT AND ORGANIZATION	Revision Date:	

1. The page borders and format should be selected and used exclusively for all manuals that will comprise company policies and procedures—and for nothing else. This way, these pages and manuals will visually assume their clear identity. One look and it is immediately apparent the page content is official company policy—with respect to safety, or to every other company issue (operations, personnel, etc.).
2. The logo placement on each page provides corporate identity and emphasizes the nature of the material firmly as company policy. It also emphasizes the proprietary nature of the material, and reinforces the employee acknowledgment signed by each employee.
3. The title block at the upper right further identifies the page as a component of an official standard company policy and procedure manual. It includes important content information (Manual Name, Manual Section, and Page Number) and the revision date of the material. It is also a good idea to provide an "Approval" section somewhere in the block to be initialed by a senior company individual. This not only clarifies that the section page is not a "draft," but also the initials of the senior individual add still another clear indication of the importance of the material as an absolute company directive with which each employee is fully expected to comply.
4. Reproduce the following sample page in full 8–1/2" × 11" size, leaving approximately 3/8" borders on the top, right, and bottom, and an approximately 3/4" border on the left. This will leave maximum room for printed material, and for hole punches for use in a three-ring binder.

Sample Safety and Loss Control Program Pages:

1. **Sample Cover Page**
2. **Sample Program Manual Identification Page**
3. **Sample Employee Acknowledgement Page**

The three sample pages that follow are designed to serve as an example of the first three pages in an actual *Construction Safety and Loss Control Program Manual.* Their use is intended as follows:

1. Sample Cover Page (see page xli)

The *Sample Cover Page* highlights that the final style of the company's actual cover page should be designed with consistency. It should be arranged in a style and format that is identical to that of all other official company programs. In this way, its appearance will immediately communicate that the manual is an official company program that communicates an organized set of standard company operating procedures that all employees will be expected to follow.

2. Sample Program Manual Identification Page (see page xliii)

The *Sample Program Manual Identification Page* is an example of the manner in which each issued manual can be kept track of. By its prominent presentation, it emphasizes to the issued employee the major points described in more detail in the *Sample Employee Acknowledgement*. It highlights the ideas that the program manual is the property of the company, that the final information is proprietary to the company operation, and that the proper attention and use of the information is a clear expectation of the issued employee.

3. Sample Employee Acknowledgement Page (see page xlv)

The *Sample Employee Acknowledgment Page* is an example of a memorandum-style notice and record. It formally transmits the final company safety and loss control program manual to the specific employee, highlights the expectations of the employee regarding the treatment and use of the material, and outlines proper use. Note that a significant value is placed on the manual, to be considered in the event that the employee is careless and/or discourteous enough to allow the manual to be abused, lost, or stolen. The sample language is actually intended to emphasize the significance that company management places on the treatment expected of the program and the program manual. Actual imposition of the reimbursement requirement will be subject to management discretion and the advice of your own human resources professionals.

The *Quick➳Disk*

The Quick➳Disk *is a 3–1/2" diskette provided with each copy of this Construction Safety & Loss Control Program Manual*. Included on the diskette are the individual sample sections as each is presented in the sample program manual. The section files are arranged to be available for reproduction by all popular DOS-based word-processing software.

If after reading the sample *Construction Safety and Loss Control Program Manual* and understanding the modification process recommended to customize the manual for use by your own company you determine that you are ready to take the next step in producing your own manual, use the contents of the Quick➳Disk to print a double-spaced copy of the particular sample sections.

Before opening the Quick➳Disk read and understand the Quick➳Disk *Copyright Notice* and *Terms of Use*. Print the desired sample sections. Use the copy as the first draft of your custom manual, and edit as with any other document. This procedure will immediately save your company hundreds of labor hours in the production of the first draft of your new *Safety and Loss Control Program Manual.*

Sample Manual

(Sample Cover Page)

(Company Name)

Safety and Loss Control Program Manual

The complete Safety and Loss Control Program Manual is the property of (**COMPANY**), and has been prepared for its exclusive use. Unauthorized reproduction, distribution, or any other use of any material herein without written permission is strictly prohibited.

Company Logo

(Sample Program Manual Identification Page)

(Company Name)

Safety and Loss Control Program Manual

Manual No.: ______________________________

Assigned to Employee:

Name: ______________________________

Number: ______________________________

Date: ______________________________

The information in this (**COMPANY**) Safety and Loss Control Program Manual is proprietary, confidential, is the property of (**COMPANY**), and has been prepared for the sole and exclusive use of (**COMPANY**) employees.

The Manual is to remain in your possession throughout your employment, and is to be maintained and kept in good condition at all times.

Intentional dissemination of the Manual's information to outside companies or to non-(**COMPANY**) individuals will be grounds for discharge. Unauthorized reproduction, distribution, or any other use of any material herein without written permission is strictly prohibited.

(Sample Employee Acknowledgement)

Safety and Loss Control Program
Employee Acknowledgment

To: Each *Safety and Loss Control Program Manual* recipient Date:

From: (**Individual Name**), Vice President

Subject: Manual receipt, use, and acknowledgment.

Please read the following and sign where indicated:

This *Safety and Loss Control Program Manual* is the property of (**COMPANY**). Its distribution is being made to (**COMPANY**) personnel for use as a reference guide. The information is proprietary, confidential, copyrighted, and for the sole and exclusive use of (**COMPANY**) employees. The Manual is to remain in your possession throughout your employment. Intentional dissemination of the Manual's information to outside companies or to non-(**COMPANY**) individuals will be dealt with the full extent allowed by law.

Do not:

- Give the Manual to anyone else.
- Reproduce it in whole or in part for use by anyone other than a (**COMPANY**) employee.
- Loan or leave it with outside parties, companies, or agencies.

Do:

- Use it to train your staff, or any other (**COMPANY**) employee as appropriate.
- Reproduce portions as appropriate *for exclusive use of (***COMPANY***) employees* who have not been issued the complete Manual.
- Make suggestions for changes, corrections, and improvements in either policy or in the way policy is communicated or applied. Such suggestions may be submitted to the (**COMPANY**) Safety & Health Manager, or to any member of (**COMPANY**) senior management.

It is the responsibility of the employee to keep the Manual current with all revisions as they are issued, and to maintain the Manual in good condition. The Manual and all revisions must be returned to the **(COMPANY)** Personnel Department as part of any exit interview, and in any event at the termination the employment relationship. *Failure to return the Manual will result in the employee reimbursing* **(THE COMPANY)** **($ INSERT VALUE)**; the full amount of which may be deducted from any final pay check otherwise due the employee.

I hereby acknowledge receipt of this Safety and Loss Control Program Manual No. ______ for my personal use and reference while an employee of **(COMPANY)**. I recognize that the information contained in the Manual is proprietary and confidential trade secrets, and is not to be released in whole or in part to companies or individuals outside **(COMPANY)**. I accept responsibility for the security of the Manual, and upon request will return the Manual to my supervisor.

Employee Name: ______________________________ No. ________________

Signature: ____________________________________ Date:_______________

Construction Safety and Loss Control Program Manual

Sample Section 1

SAFETY AND LOSS CONTROL PROGRAM: CONTENT AND ORGANIZATION

Section Use

The form and content of Sample Section 1 *Safety and Loss Control Program: Content and Organization* is designed as an example of company instruction to its employees with respect to the material presented. It is intended to provide for each employee:

- Explanation of the overall company safety and loss control program;
- Basic instruction regarding the manner in which the *Safety and Loss Control Program Manual* relates to the complete company program;
- Notice to each employee that all material contained in the program manual has been prepared for the exclusive use by the company; and
- Description of the manner in which the *Safety and Loss Control Program Manual* correlates with other printed company information.

The sample section that follows is an example of a company safety and loss control Program approach to the objectives indicated above. The sample language and forms are to be considered in the context of your own company and its manner of operation.

Carefully consider these objectives during the review and editing process of the section material in order to customize the sample section content to the specific needs of your organization. Prepare final language that is consistent with decisions made by your own management in the determination of your own company program.

It is important to understand that safety rules and regulations change from time to time. Be sure to review the final draft of the section and the entire program manual with appropriate safety, insurance, and legal professionals prior to issue and use. Periodically review your program with company management and all appropriate professionals to be certain that the company program remains current with all requirements.

COMPANY LOGO	Construction Safety and Loss Control Program Manual	Page: 1-1	Appr:
	1. CONTENT AND ORGANIZATION	Revision Date:	

(Sample)

SECTION 1

Safety and Loss Control Program Structure

COMPANY LOGO	Construction Safety and Loss Control Program Manual	Page: 1-2	Appr:
	1. CONTENT AND ORGANIZATION	Revision Date:	

1.1 PURPOSE

NOTE TO MANUAL PREPARER: This sample section assumes a specific organizational structure of your company, and a particular approach to organizing and communicating your own safety and loss control objectives. Consider this sample section carefully in this context, and modify the material to accommodate the actual manner in which your own company intends to deal with the issues presented.

The purpose of this ***Safety and Loss Control Program*** is to provide for the systematic identification, evaluation, and prevention or control of general workplace hazards, specific job hazards, and potential hazards that may arise from foreseeable conditions.

This ***Safety & Loss Control Program Manual*** has been prepared so that each (**COMPANY**) employee will have a reference as to what is expected of him or her with regard to safe working practices on each (**COMPANY**) jobsite and shop area. It must be clear that safety on each site and in each area is our *highest priority*. We maintain a Safety Committee, provide ongoing safety training for (**COMPANY**) employees, and employ a Safety and Health Manager to monitor and assist our employees and contribute to safety training.

Employee rights and responsibilities go hand in hand. We fully expect that no (**COMPANY**) employee will ever intentionally expose any worker to an unsafe condition, or observe actual or possible exposure without reporting the instance to his or her supervisor as soon as possible. Employees are expected to develop their own commitment to safety and health protection for themselves and for their fellow workers. Any employee who does not comply with the company safety policy, or who is persistently negligent in his or her responsibilities in this regard may be subject to disciplinary action or dismissal.

1.2 SCOPE AND INTENT

The *Safety and Loss Control* Program establishes the duties and responsibilities of the field supervisory staff, as well as the basic procedures to be followed by each (**COMPANY**) employee. The regulations and objectives are to help assure the health and safety of each (**COMPANY**) employee, their fellow workers, all persons authorized to be on each jobsite, and the public.

In the preparation of this Manual, every effort has been made to be complete, yet practical. It is the explicit intention of (**COMPANY**) that each program, policy, and procedure described herein be comprehensive, and comply with all applicable current laws to the best of our ability to interpret and apply them to our daily work processes.

It is possible that a process or requirement on any jobsite may not be completely addressed, or that a new process or requirement may be encountered. In such conditions, it is understood that each new situation must be addressed by immediately referring to and applying local, state, and/or federal laws governing the situation—*prior to performing the related work activities.* If there is any question, contact the **(COMPANY)** Safety and Health Manager for assistance and direction. The Safety and Health Manager has the authority to halt any work activity which he or she believes may be an imminent threat to life, health, property, or equipment.

1.3 MANUAL ORGANIZATION

The Manual is divided into major sections for clarity and ease of reference. Each major section includes the **(COMPANY)** company policies and procedures with respect to all project and employee requirements, responsibilities, and expectations.

The ***Safety and Loss Control Program Manual*** is further coordinated with three other company procedure manuals:

1. The ***Jobsite Safety and Equipment Safety and Maintenance Manual,***
2. The ***Hazard Communication Program Manual,*** and
3. The ***Human Resources Manual***

These manuals are explained in their respective sections in this Manual. Together, all four manuals make up the complete ***(COMPANY) Safety & Loss Control Program***, and should be maintained as a single set.

NOTE TO MANUAL PREPARER: Some of the more comprehensive company safety and loss control programs may include separate manuals for distribution of relevant material to individuals and jobsites, thereby avoiding unnecessary or undesired duplication and dissemination of the complete program. In addition, the safety and loss control information is typically coordinated with other materials relevant to the human resources management effort. This sample Safety and Loss Control Program Manual anticipates such an arrangement. The three manuals references to are not included in this Safety and Loss Control Program Manual, but should comprise portions of the company's entire set of organized policies and procedures.

The Jobsite Safety and Equipment Safety and Maintenance Manual (see

Section 1.3.1), the Hazard Communication Program Manual (see Section 1.3.2), and the Employee Booklet (see Section 1.3.4) may be prepared as separate manuals, and coordinated with your actual company Safety and Loss Control Program Manual in its final form.

Similarly the Human Resources Manual (see Section 1.3.3) is not included herein but is referenced to highlight the importance of coordinating the final Safety & Loss Control Program with all company human resource procedures.

1.3.1 Jobsite Safety and Equipment Safety and Maintenance Manual

All (**COMPANY**) foreman, superintendents, and equipment operators are provided with their own copy of the ***(COMPANY) Jobsite Safety and Equipment Safety and Maintenance Manual.***

(**COMPANY**) policy and specific instruction are provided in the form of direct communication from employer to employee, in order to provide the information most relevant to daily operations. Also included is the identification of this complete *Safety and Loss Control Program Manual*, the role of supervisors with regard to dissemination of safety information, and the availability of the complete Manual for their review and reference.

The Jobsite and Equipment Safety and Maintenance Manual is a complete reproduction of Sample Sections 15 and 16 of this *Safety and Loss Control Program Manual.* It is arranged as a separate field manual in order to ensure that the complete set of field and equipment guidelines are provided to each employee who requires the information on a daily basis, but who may not need to have the complete *Safety and Loss Control Program Manual* at the jobsite.

1.3.2 Hazard Communication Program Manual

The complete ***Safety and Loss Control Program Manual*** will further be coordinated with the separate ***Hazard Communication Program Manual (HAZ-COM Manual)*** described in Section 9 of this Manual.

The HAZ-COM Manual is a *separate book* containing the actual MSDS (Material Safety Data Sheets) for those materials commonly used by (**COMPANY**) on its jobsites.

1.3.3 Human Resources Manual

The (**COMPANY**) ***Human Resources Manual*** describes the complete set of company policies and procedures with respect to the human resources function. It includes the (**COMPANY**) ***Managed Care / Drug-Free Workplace Program,*** and the **(COMPANY*)*** ***Substance Abuse Program***, and is further coordinated with the related safety policies contained in this ***Safety and Loss Control Program Manual***.

NOTE TO MANUAL PREPARER: The Managed Care/Drug-Free Workplace Program referred to above is an example of certain specific programs that may be available from your insurance professionals. Such programs may offer discounts on insurance costs for very specific programs and conduct established and maintained by the company. Check with your own insurance provider for the advisability and availability of such programs.

If such a program is selected, modify this sample section to properly incorporate the program in the complete **Safety & Loss Control Program.**

1.3.4 Employee Booklet

Each (**COMPANY**) employee is given upon hire the ***(COMPANY) Employee Booklet*** more completely described in the complete (**COMPANY**) Human Resources Manual. General guidelines and instruction to all employees with respect to their safety rights and responsibilities are included therein, along with the identification of this Safety & Loss Control Program, the role of their supervisors with regard to dissemination of safety information, and the availability of the complete Manual for their review and reference.

1.4 (COMPANY) SAFETY AND HEALTH MANAGER

NOTE TO MANUAL PREPARER: This subsection requires identification of a company employee who is charged with the primary responsibility of a Safety and Health Manager as described in this Sample Section.

The Company Safety and Health Manager is:

(Individual's Name)
Mobile Phone: (000) 000–0000
Pager: (000) 000–0000

This individual is the authorized company official who will be available to answer questions and disseminate information with respect to company safety policies, programs, and procedures. Contact him (her) whenever necessary.

COMPANY LOGO	Construction Safety and Loss Control Program Manual	Page: 1-6	Appr:
	1. CONTENT AND ORGANIZATION	Revision Date:	

1.5 MANUAL ISSUE, REVISIONS, AND USE

The Manual is issued to all (**COMPANY**) supervisory personnel, including Project Managers and Site Superintendents. Through these individuals, its information is to be made available to all (**COMPANY**) employees to refer to at any time. As future needs and changes in requirements occur, revisions will be made, and each employee will be advised.

This ***Safety and Loss Control Program Manual*** is the property of (**COMPANY**). It is the responsibility of the issued employee to keep the Manual current with all revisions as they are distributed, to maintain the Manual in good and usable condition at all times, and to return the Manual to (**COMPANY**) upon termination of employment.

Any clarifications or variances sought must be obtained through the Company Safety and Health Manager.

1.6 NEW HIRE TRAINING

Every new hire at (**COMPANY**) will receive a personal condensed version of this safety manual. He or she will be instructed to read through it prior to beginning work, and directed to keep the personal manual available at all times. The Personnel Director or the Safety and Health Manager will conduct a safety orientation with the new hire(s), concluding with a short time to answer any questions that the new hire(s) may have concerning safety. At the end of this orientation session, the new hire(s) may be released to his or her supervisor for assignment.

Rehires of less than six (6) months duration need not be considered new hires for the purposes of safety orientation.

1.7 SAFETY ORIENTATION VIDEO

NOTE TO MANUAL PREPARER: This sample subsection applies only if the company has made the effort to prepare its own video or to secure such a video from an alternative source (such as a company providing such a service).

As part of the employment orientation process described in Section 1.6, each new (**COMPANY**) employee will be required to view the (**COMPANY**) Safety &

Employment Orientation Video. The visual program:

- Introduces key (**COMPANY**) individuals involved in the Safety, Loss Control, and Human Resources Policies and Objectives of the Organization;
- Provides an overview of the program, the procedural processes, and the use of various important forms;
- Summarizes employment responsibilities;
- Identifies key individuals within the (**COMPANY**) organization, and their responsibilities with respect to a number of safety and employment issues; and
- Remains as a reference tool for all (**COMPANY**) employees to refer to whenever desired.

Sample Section 2

SAFETY AND HEALTH POLICY

Section Use

The form and content of Sample Section 2 *Safety and Health Policy* is designed as an example of company instruction to its employees with respect to the material presented. It is intended to provide for each employee:

- A clear statement of company policy with respect to safety;
- A statement of company safety objectives;
- A statement of company expectation of each employee regarding their conduct in all company environments;
- A summary of disciplinary action for violating any company safety rules and policies.

The sample section and forms that follow are examples of a company safety and loss control program approach to the objectives indicated above. The sample language and forms are to be considered in the context of your own company and its manner of operation.

Carefully consider these objectives during the review and editing process of the section material in order to customize the sample section content to the specific needs of your organization. Prepare final language that is consistent with decisions made by your own management in the determination of your own company program.

It is important to understand that safety rules and regulations change from time to time. Be sure to review the final draft of the section and the entire program manual with appropriate safety, insurance, and legal professionals prior to issue and use. Periodically review your program with company management and all appropriate professionals to be certain that the company program remains current with all requirements.

COMPANY LOGO	Construction Safety and Loss Control Program Manual	Page: 2-1	Appr:
	2. SAFETY AND HEALTH POLICY	Revision Date:	

(Sample)

SECTION 2

Safety and Health Policy

COMPANY LOGO	Construction Safety and Loss Control Program Manual	Page: 2-2	Appr:
	2. SAFETY AND HEALTH POLICY	Revision Date:	

2.1 COMPANY SAFETY POLICY

As (**COMPANY**) continues through our (*number*) year of providing construction services, our consistent record of safety, and our concern for the safety, health, and welfare of all of our employees has become our greatest asset. We must all continue to recognize that there is no other aspect of our work that takes on greater priority.

It is the policy of (**COMPANY**) to:

- Provide safe working environments, and to perform all company activities in a manner that reduces risk to all employees and to all other workers;
- Maintain all office, equipment, and jobsite conditions in ways that eliminate risk to visitors and the public, and eliminate risk of damage to property and equipment;
- Provide safe, well-maintained equipment, proper training for all equipment operators, and instruction on safe methods and procedures;
- Comply with all federal, state, and local laws and regulations as they apply to all work performed;
- Never accept any unsafe working condition for any reason, and to take immediate corrective action when any safety violation is observed.

Correspondingly, we expect that:

- Each *employee* of (**COMPANY**) will be responsible for safety and health;
- All safety equipment issued by (**COMPANY**) must always be used as intended;
- Company safety and all other equipment must not be damaged, removed, altered, or abused, or operated when observed to be in such condition;
- Employees must be familiar with and observe all safety rules, procedures, and policies, as they may change from time to time;
- Supervisors will see that all rules and procedures are observed by their crews, and immediately enforce appropriate corrective measures whenever violations are observed.

The collective results of all our direct attention to the safety objective will contribute to company success and the pride and security that goes with it. Conduct with respect to safety will affect the manner in which the performance of all employees will be measured.

Although we enjoy a safety record to be proud of, *our goal is 100% accident-free work, while ensuring our history of enduring quality work and satisfied clients*. The good intentions, cooperation and good judgment of all **(COMPANY)** employees in the use of safe and responsible work practices is the path toward continued personal and company improvement, and must be pursued each day.

This ***Safety and Loss Control Program Manual*** defines respective company and employee responsibilities, participation in the program, and individual activities to be performed by each. The coordinated, cooperative approach will minimize labor and monetary losses, and help all **(COMPANY)** employees to comply with all current laws and regulations.

2.2 JOBSITE SAFETY POLICY NOTICE

During initial project mobilization, the notice on the following page is to be reproduced on **(COMPANY)** letterhead and placed in a prominent location at the field office, in any bulletin area provided on a jobsite for such notices, or in any other appropriate location. It is a statement of **(COMPANY)** Company Policy, and of the conduct expected of every person on the site, including employees and non-employees.

Sample Jobsite Safety Policy Notice:

(Sample)
Jobsite Safety Policy

It is the Policy of **(COMPANY)** to at all times:

- Provide safe working conditions,
- Perform all activities in ways that eliminate risk of injury or health impairment to any tradesperson,
- Maintain all areas in ways that eliminate risk to visitors and to the public, and
- Eliminate risk of damage to property on and adjacent to every jobsite.

These are basic responsibilities of every company and individual on every jobsite. All supervisors of all trades must routinely accept complete responsibility for prevention of accidents and for the safety of all work under their direction. All tradespeople of every category are required to conduct themselves in a safe, considerate, and workmanlike manner.

By contract and by law, every company and person employed on the site is obligated at a minimum to comply with the Federal Occupational Safety and Health Act, Americans with Disabilities Act, and the laws of every entity having jurisdiction over the work and the site.

Any company or individual refusing to correct observed safety violations will be banned from the site at least until such violations are corrected, and will be held completely responsible for all resulting effects.

With attention, awareness, and cooperation by everyone, we will all achieve an accident-free job, and the pride that goes with it.

(COMPANY)

(Individual Name), President (***or other company officer***)

2.3 SUBSTANCE ABUSE AND (COMPANY) DRUG-FREE POLICY

NOTE TO MANUAL PREPARER: If the company does not currently have an Employee Handbook, correct the references included below.

The **(COMPANY)** Substance Abuse Policy is contained in the **(COMPANY)** Employee Handbook that is distributed to all employees and new hires.

2.4 DISCIPLINARY ACTION AND PROCEDURE FOR SAFETY VIOLATIONS

2.4.1 Policy

Compliance with all company safety rules and procedures is a condition of employment when working for **(COMPANY)**. All employees must familiarize themselves with company safety rules and procedures, and comply with them in every respect. Supervisory, administrative, and management personnel at all levels are responsible for taking immediate corrective action when a violation is observed. Foremen are responsible for their crews' compliances.

Any employee causing or knowingly allowing an unsafe condition to remain shall be subject to a warning and possible dismissal. Employees guilty of intentional serious and/or repeated violations will be subject to dismissal without pay.

2.4.2 Disciplinary Action

If a violation is observed, or comes to the attention of any supervisor, administrator, or management personnel, action must be taken immediately to correct the violation. Immediately thereafter, the Safety and Health Manager is to be notified. The Safety and Health Manager will then follow the procedures below for necessary disciplinary action:

FIRST WARNING:
The first warning will be verbal and written, with a copy of the ***Safety Violation Warning Notice*** given to the employee and distributed to the employee's personnel record. Refer to Sample Section 2.4.3 of this *Safety and Loss Control Program Manual.*

SECOND WARNING:
A written warning similar to the first warning will be given to the employee, be retained by the Safety and Health Manager, and be distributed to the employee's personnel record. A meeting will be held with the employee, foreman, and Safety and Health Manager in order to determine why the employee is not willing to comply with the rules and regulations of the company. Any action taken at this time will be determined by management, and be based upon the severity of the violation.

THIRD WARNING:
A written warning will be prepared and a copy will be given to the employee, be retained by the Safety and Health Manager, and be distributed to the employee's personnel record. Three (3) warnings for safety violations within a twelve (12) month period will result in suspension or termination of employment. Three (3) safety violations with a six (6) month period will result in termination of employment without pay.

The actions listed above must be taken when a violation is observed. (**COMPANY**) cannot tolerate actions or negligence that may result in an injury. If there are any questions concerning this policy and procedure, contact the Safety and Health Manager.

2.4.3 Safety Violation Warning Notice

The ***Safety Violation Warning Notice*** on the following page is to be used as prescribed in the preceding section, whenever a safety violation is observed:

COMPANY LOGO	Construction Safety and Loss Control Program Manual	Page: 2-6	Appr:
	2. SAFETY AND HEALTH POLICY	Revision Date:	

(COMPANY) Safety Violation Warning Notice

Notice Date: ____________

Employee Name: ____ #: ______________

Date of Violation: ____________________ Time of Violation: ________:_____ AM PM

Location: ____

Description of Violation: ____

____

____

____

____

____

____

Course of Action: ____

____

____

____

____

____

____

Employee Warned of: ☐ Suspension ☐ Dismissal ☐ Dismissal Without Pay

Signed:

Safety & Health Mgr.: _______________________ Employee: _______________________

cc: ☐ Supervisor
☐ Employee
☐ Personnel

Sample Section 3

SAFETY AND LOSS CONTROL EMPLOYEE PARTICIPATION

Section Use

The form and content of Sample Section 3 *Safety and Loss Control Employee Participation* is designed as an example of company instruction to its employees with respect to the material presented. It is intended to provide for each employee:

- The function of the company Safety Committee;
- The participation of each management level and of every employee, and how their efforts are coordinated with each other; and
- The responsibilities of each employee, and the conduct expected by the company.

The sample section that follows is an example of a company safety and loss control program approach to the objectives indicated above. The sample language and forms are to be considered in the context of your own company and its manner of operation. The sample section anticipates a specific organization and staff arrangement within the company, as well as a precise arrangement of a Safety Committee. The manual preparer must review the entire sample section very carefully in consideration of the intentions of the company to provide for the management, committee, and staff arrangements as discussed. The final actual section must then be written to accommodate the specific arrangements ultimately determined by your own company management.

Carefully consider these objectives during the review and editing process of the sample section material in order to customize the section content to the specific needs of your organization. Prepare final language that is consistent with decisions made by your own management in the determination of your own company program.

It is important to understand that safety rules and regulations change from time to time. Be sure to review the final draft of the section and the entire program manual with appropriate safety, insurance, and legal professionals prior to issue and use. Periodically review your program with company management and all appropriate professionals to be certain that the company program remains current with all requirements.

COMPANY LOGO	Construction Safety and Loss Control Program Manual	Page: 3-1	Appr:
	3. SAFETY AND LOSS CONTROL EMPLOYEE PARTICIPATION	Revision Date:	

(Sample)

SECTION 3

Safety & Loss Control Employee Participation

COMPANY LOGO	Construction Safety and Loss Control Program Manual	Page: 3-2	Appr:
	3. SAFETY AND LOSS CONTROL EMPLOYEE PARTICIPATION	Revision Date:	

3.1 RESPONSIBILITY ASSIGNMENT

Whether at the jobsite, warehouse, shop, or office, every employee has a responsibility to themselves, to the other employees, and to the company to work safely. That responsibility includes the elimination of hazards wherever possible, and reporting actual or potential hazards to management or to the Safety and Health Manager immediately upon their observation.

Supervisors are held accountable for compliance with all safety objectives in the performance of every activity. This is especially so in the areas of training and hazard control. Any performance evaluation(s) will reflect their involvement and active participation in accident prevention.

The fact that the company has assigned a Safety and Health Manager does not release managers or supervisors from their safety and health responsibilities in any way. The Safety and Health Manager will be available to assist supervisors in implementing the company Safety and Health Program, but will not be expected to perform the respective supervisors' duties.

3.2 THE (COMPANY) SAFETY COMMITTEE

3.2.1 Description

The (**COMPANY**) Safety Committee is made up of both employer and employee representatives who are charged with the responsibility of general oversight of the (**COMPANY**) Safety & Loss Control Program. The Committee convenes quarterly, and performs the important functions described in Section 3.2.2.

3.2.2 Committee Function and Objectives

1. Meet regularly to review the overall Safety & Loss Control Program, and specific operating issues that may arise during the period.
2. Serve in an advisory capacity to the Safety and Health Manager, and to management.
3. Familiarize themselves with applicable construction safety standards.
4. Review established company procedures and evaluate the effectiveness of their implementation.
5. Recommend corrections and improvements in procedures, safety rules, and

company policies with respect to accident and illness prevention on our jobsites and in our offices.

6. Communicate procedures approved by **(COMPANY)** management to all **(COMPANY)** employees. Ensure that all updates and changes to company procedure as approved by **(COMPANY)** management are adopted, properly coordinated with the entire company Safety & Loss Control Program, and distributed.
7. Participate in communication procedures by which **(COMPANY)** shall train committee members and **(COMPANY)** employees.
8. Prepare and distribute minutes of committee meetings. Make records of committee activities and communications available to **(COMPANY)** employees.

3.2.3 Safety Committee Structure and Duties

SAFETY COMMITTEE DUTIES:

- Develop company-wide action plans for compliance with safety regulations.
- Develop an action program for project accountability.
- Provide suggestions and direction for efficient implementation of prescribed corrective measures.
- Direct the distribution of safety regulations, programs, and other informational materials.

CHAIRMAN DUTIES:

- Schedule and enforce participation in all Safety Committee meetings.
- Prepare and distribute Safety Committee meeting agendas.
- Report to senior company management regarding the status of all recommendations.
- Ensure preparation of minutes of meetings, and distribute.
- In the absence of the Chairman, the Co-Chairman is authorized to assume all responsibilities and authority of the Chairman.

MEMBER DUTIES: ***(Superintendents, Foremen, Managers, and Employees)***

- Participate in all Safety Meetings.
- Assist with safety inspections.

- Apply and enforce OSHA and company regulations.
- Report unsafe conditions.
- Report all accidents or near accidents.
- Contribute ideas and suggestions for improvement of safety and better communication of safety programs and information.

3.3 THE SAFETY AND HEALTH MANAGER

NOTE TO MANUAL PREPARER: The Safety and Health Manager is intended in this sample Safety and Loss Control Program Manual to be an individual in a larger organization who has the dedicated function as described. In a large company, the function may require one or more full-time individuals. In small to mid-sized companies, the function may not require (or allow) a separate full-time individual, but necessarily may require that the responsibilities be either provided for collectively by company project management and site superintendence, or given as part-time responsibilities to an individual.

Consider these arrangements very carefully, determine the specific arrangement to be adopted by your own company, and rewrite this section to distribute responsibilities according to the structure adopted.

The company Safety and Health Manager is responsible for the administration of routine dissemination of all company Safety & Loss Control information, for the administration of all company Safety & Loss Control procedures, and for monitoring the compliance of **(COMPANY)** employees with all stated policies and procedures. The Safety and Health Manager is to be recognized as a deputy of each Senior Project Manager, and shall report directly to them. In matters of imminent safety concern, the Safety and Health Manager will simultaneously report directly to company senior management. Any recommendations issued by the Safety and Health Manager with respect to safety and loss control are to be considered as issued by the Senior Project Manager.

IN SITUATIONS WHERE IMMINENT DANGER OR SERIOUS HAZARD EXISTS, THE SAFETY AND HEALTH MANAGER HAS THE AUTHORITY TO ORDER A WORK SUSPENSION.

The Safety and Health Manager will assist the Project Managers and Site Superintendents in all matters pertaining to safety and loss control. Duties of the Safety and Health Manager will include but may not be limited to:

COMPANY LOGO	Construction Safety and Loss Control Program Manual	Page: 3-5	Appr:
	3. SAFETY AND LOSS CONTROL EMPLOYEE PARTICIPATION	Revision Date:	

- Assist each company employee with compliance with all company and OSHA safety and health policies and regulations.
- Implement the company Safety, Health, & Loss Control Programs, and monitor compliances.
- Coordinate all company safety activities in ways that facilitate their implementation.
- Advise on the purchase of safety and health materials to ensure compliance with all safety standards.
- Advise management regarding proposed and/or necessary changes in safety standards and regulations.
- Conduct field inspections in efforts to identify unsafe conditions and/or actions of jobsite personnel. Make verbal and written recommendations for both immediate and future correction. Follow up to assure correction. Issue warnings of persistent, uncorrected unsafe conditions.
- Organize and conduct training of supervisory and hourly employees in safe work procedures.
- Be thoroughly familiar with applicable state, local, and OSHA laws, procedures, and standards. Be available to assist management in interpretations, development of policy, and implementation of all adopted procedures.
- Represent (**COMPANY**) management during any local, state, insurance, or OSHA inspections in every jobsite situation.
- Coordinate the deployment of emergency care systems, such as first aid, medical, fire protection, evacuation, and fire alarms.
- Be knowledgeable of health-related and hygiene-related activities, and pursue continuing education with respect to same.
- Coordinate any/all site security procedures and personnel, when required for the specific project.
- Assist all Project Managers in the implementation of the (**COMPANY**) safety programs as they relate to the specific projects. Maintain communication with (**COMPANY**) management.
- Respond to company managers and supervisors with answers to their requests for information.
- Coordinate jobsite meetings with (**COMPANY**) Project Managers. Assist or lead weekly jobsite Tailgate Safety Meetings for (**COMPANY**) trades to the extent available to do so. Assist all project managers with organization and presentation of weekly Tailgate Safety Meetings.

COMPANY LOGO	Construction Safety and Loss Control Program Manual	Page: 3-6	Appr:
	3. SAFETY AND LOSS CONTROL EMPLOYEE PARTICIPATION	Revision Date:	

- Assist the Project Managers in the establishment and maintenance of all jobsite notices, signs, etc., and maintenance of each "# Days Safely Worked" sign on each company jobsite.
- Coordinate the organization and implementation of all emergency care warning and evacuation systems.
- Maintain all accident reporting and documentation procedures, including preparation of all reports, notifications to all parties, and accident investigation reporting.
- Promptly notify **(COMPANY)** management and other required parties whenever necessary.
- Assist in the preparation and presentation of the monthly Supervisory Safety Meeting, and quarterly Employee Safety Meetings.
- Coordinate and participate in disciplinary actions and procedures.
- Review all accident reports for correct form, complete information, and proper preparation.
- Periodically evaluate the effectiveness of all Safety, Health, & Loss Control Programs, and make appropriate recommendations to management.

3.4 THE SENIOR PROJECT MANAGER

The Senior Project Manager is responsible for supervising and monitoring the Safety & Loss Control Programs for all projects under his or her direct supervision, and through each Project Manager for those projects being reported to him or her. Safety inspections are to be conducted by the Senior Project Manager during all routine jobsite tours.

3.5 THE PROJECT MANAGER

The Project Manager is ultimately responsible for the complete Safety & Loss Control effort on the projects under his or her direct authority. All reporting will be made to the Project Manager, who will:

- Plan production so that all work will be performed in accordance with established safety regulations.

- Administer the Safety & Loss Control Program on all projects under his or her control.
- Disseminate all safety information to all appropriate personnel and subcontractors.
- Receive and process all reports, surveys, accident reports, and other information relating to safety and loss control that are to be submitted to the Project Manager.
- Determine the need for corrections that may be necessary to remedy or improve various situations, and ensure that such action is implemented and maintained.

The Project Manager shall coordinate his or her specific responsibilities as listed in this Section 3.5 with the Site Superintendent and Safety and Health Manager, and determine the most effective implementation of all Safety & Loss Control responsibilities between the Project Manager and Site Superintendent.

3.6 THE SITE SUPERINTENDENT

The Site Superintendent, as coordinated with and under the direction of the Project Manager, is responsible for components of all aspects of the jobsite safety program, and for the proactive administration of all jobsite safety and loss control activities. The responsibilities of the Site Superintendent may be coordinated with and/or shared by the Project Manager as the Project Manager will determine, and in any event will include but may not be limited to:

- Plan production so that all work will be performed in accordance with established safety laws, procedures, and regulations.
- Assist in the preparation, notification, and monitoring of the Jobsite Safety Program.
- Establish, post, and maintain all jobsite safety and health notices, signs, etc.
- Regularly conduct jobsite and work area inspections in an effort to identify unsafe or potentially unsafe conditions and/or activities of jobsite personnel.
- Make recommendations for both immediate correction and future improvements in each condition. Follow through on each recommendation in order to ensure proper responses and continuing improvements.
- Be knowledgeable of all applicable local, state, and OSHA rules, procedures, regulations, and standards.

COMPANY LOGO	Construction Safety and Loss Control Program Manual	Page: 3-8	Appr:
	3. SAFETY AND LOSS CONTROL EMPLOYEE PARTICIPATION	Revision Date:	

- Be prepared and available to accompany any local, state, insurance, or OSHA inspectors to represent (**COMPANY**) management and/or assist the Safety and Health Manager or project manager in such representation whenever requested to do so.
- Coordinate the organization and development of emergency care materials, including first aid equipment, medical materials, and fire protection equipment.
- Coordinate emergency evacuation procedures and fire alarm provisions with the company Safety and Health Manager.
- Be familiar with health issues, and or industrial hygiene. Be aware of all procedures needed to deal with industrial hygiene issues.
- Coordinate site security procedures and personnel as may be required for the specific project, and as may be determined to be necessary by the project manager.
- Lead weekly jobsite Tailgate Safety Meetings for (**COMPANY**) trades. Assist the Safety and Health Manager in the presentation of Tailgate Safety Meetings when he or she is available for such meetings as coordinated.
- Coordinate emergency care warning and evacuation systems as required for the specific project.
- Ensure that proper safety material, equipment, and protective devices are available, well maintained, in correct working order, and used whenever required.
- Instruct all foremen in safety requirements and make certain that each foreman regularly passes the instruction on to their crews.
- Take advantage of all safety training as it is offered, be aware of all safety rules, and respect the need for continuing safety education.
- Review all accidents, oversee correction of all unsafe practices, and file accident reports.
- Conduct jobsite safety meetings as described in the related section(s) of this *Safety and Loss Control Program Manual*, and provide all employees with proper instruction on safety requirements.
- Require conformance to safety standards by all subcontractors as required by the standard company Subcontractor Safety Policy.
- Notify the (**COMPANY**) home office of all actual or alleged safety violations by any/all parties.

- Provide for the protection of the public from company operations.
- Regularly attempt to ensure safe performance and enforce all compliances by others present on the site, including the project owner and architect/engineer representatives, the general public, visitors, and the employees of other contractors.

3.7 THE JOB FOREMAN

The Job Foreman, as coordinated with and under the direction of the Site Superintendent, is responsible for the proactive administration, control, and enforcement of all provisions of the company jobsite safety program as they relate to the crews under their direction.

The responsibilities of the project foreman will include but may not be limited to:

- Carry out the Jobsite Safety Program at the work level.
- Be aware of all safety requirements and safe working practices.
- Plan all work activities in ways that comply with safe working practices.
- Instruct new employees and existing employees performing new tasks in safe working practices.
- Provide crews with continuing instruction on safety requirements, and conduct regular Tailgate Safety Meetings.
- Install and maintain all devices in a manner coordinated with the Site Superintendent and Project Manager, in order to protect the public from company operations.
- Make certain that protection equipment is available, maintained in operating condition, and used.
- Act without delay to correct all hazards including unsafe acts and conditions which are within the scope of your crews' work.
- Act to secure prompt medical attention for any injured persons.
- Immediately report all actual, perceived, or alleged safety violations to your immediate supervisor.

COMPANY LOGO	Construction Safety and Loss Control Program Manual	Page: 3-10	Appr:
	3. SAFETY AND LOSS CONTROL EMPLOYEE PARTICIPATION	Revision Date:	

3.8 (COMPANY) EMPLOYEES

Each (**COMPANY**) employee is to make themselves aware of their safety, health, and loss control responsibilities. Each employee will be expected to comply with all safety, health, and loss control rules, regulations, and stated company policies which are applicable to their work. Employees are expected to report observed safety and health violations, suspected violations, and anticipated hazards to his or her immediate supervisor immediately upon observation or detection. If such reported conditions are not subsequently corrected, the employee is to report the failed correction to the Safety and Health Manager.

Workers Shall:

- Work safely and in such a manner as to ensure their own safety, as well as the safety of coworkers and all others.
- Request help when unsure of how to perform any task safely.
- Correct unsafe acts or conditions within the scope of immediate work.
- Report any uncorrected unsafe acts or conditions to supervision.
- Report for work in good physical and mental condition to safely carry out assigned duties.
- Avail themselves of company and industry sponsored safety programs.
- Use and maintain all safety devices provided.
- Maintain and properly use all tools under his or her control.
- Follow all safety rules.
- Provide fellow employees with help in implementing safety procedures and complying with safety requirements.

All Personnel Shall:

- Strive to make all work environments and company operations safe.
- Maintain mental and physical health conducive to working safely.
- Keep all work areas clean and free of debris and obstacles.
- Assess results of their actions on the entire workplace.

COMPANY LOGO	Construction Safety and Loss Control Program Manual	Page: 3-11	Appr:
	3. SAFETY AND LOSS CONTROL EMPLOYEE PARTICIPATION	Revision Date:	

- Replace or repair safety precautions removed or altered before leaving work area.
- Abide by all safety rules and regulations of the owner and every legal authority on each jobsite.
- Work in strict conformance with OSHA regulations.

No **(COMPANY)** employee shall be required or knowingly be permitted to work in an unsafe environment, except for the purpose of making safety corrections; and then only after proper precautions have been taken for their protection.

3.9 (COMPANY) SUBCONTRACTORS AND TRADE CONTRACTORS

Each **(COMPANY)** subcontractor or trade contractor is completely responsible for the safety, health, and security of the individuals in their employment. Each is required to comply with the requirements of the **(COMPANY)** Safety Program as incorporated into their respective subcontracts or purchase agreements, and with all laws, regulations, and requirements of OSHA, the owner, and every other agency having jurisdiction over the work.

Each **(COMPANY)** subcontractor or trade contractor is to be required to submit to the **(COMPANY)** Project Manager copies of their respective safety programs prior to the start of their work. Each subcontractor or trade contractor will:

- Abide by all company and project safety rules.
- Participate in all periodic jobsite safety meetings and regular site safety inspections.
- Appoint one representative and one alternate representative with the responsibility to attend safety meetings, respond to notified safety violations, and otherwise participate in the project ongoing safety program.
- Provide an adequate first aid kit for use by its employees.
- Provide a list of employees qualified in first aid who may be available to administer first aid.
- Notify all other contractors when their activities could possibly affect the health or safety of employees of other companies.
- Report all injuries or any unsafe conditions that may become apparent to the **(COMPANY)** Project Superintendent or Project Manager.

COMPANY LOGO	Construction Safety and Loss Control Program Manual	Page: 3-12	Appr:
	3. SAFETY AND LOSS CONTROL EMPLOYEE PARTICIPATION	Revision Date:	

3.10 ARCHITECTS, ENGINEERS, OWNERS, AND VISITORS

Prior to being allowed to proceed anywhere on a jobsite, all personnel and representatives of architects, engineers, owners, and visitors shall be required to:

- Abide by all company and project safety rules.
- Check in with the project site superintendent so that possession and use of proper personal protective equipment may be confirmed or provided, such as hard hats, or eye or respirator protection.
- Refrain from entering any construction area without the knowledge of (**COMPANY**) jobsite supervision, and confirmation that all required and appropriate precautions and protections have been made.

Sample Section 4

Jobsite Safety and Health Program

Section Use

The form and content of sample Section 4 *Jobsite Safety and Health Program* is designed as an example of company instruction to its employees with respect to the material presented. It is intended to provide for each employee information regarding:

- Jobsite safety planning.
- Jobsite safety meetings.
- Jobsite safety inspections.
- Provisions for basic jobsite first aid materials and training

The sample section that follows is an example of a company safety and loss control program approach to the objectives indicated above. The sample language and forms are to be considered in the context of your own company and its manner of operation.

Carefully consider these objectives during the review and editing process of the section material in order to customize the section content to the specific needs of your organization. Prepare final language that is consistent with decisions made by your own management in the determination of your own company program.

It is important to understand that safety rules and regulations change from time to time. Be sure to review the final draft of the section and the entire program manual with appropriate safety, insurance, and legal professionals prior to issue and use. Periodically review your program with company management and all appropriate professionals to be certain that the company program remains current with all requirements.

COMPANY LOGO	Construction Safety and Loss Control Program Manual	Page: 4-1	Appr:
	4. JOBSITE SAFETY AND HEALTH PROGRAM	Revision Date:	

(Sample)

SECTION 4

Jobsite Safety and Health Program

4.1 SAFETY AND HEALTH ADMINISTRATION

Each worksite must be regularly analyzed on a continuing basis in order to identify existing actual or potentially hazardous conditions. It is the responsibility of the Project Manager to administer the complete Jobsite Safety Program through the Site Superintendent and in a manner coordinated with the Safety and Health Manager.

The consideration of the Safety Program begins at the site utilization planning stage, and continues throughout the life of the project. Consider the Site Safety Planning Checklist of Section 4.3 as a starting point to help organize the particular project's Jobsite Safety Program.

4.2 JOBSITE SAFETY AND HEALTH INSPECTION PROGRAM

4.2.1 Description

(**COMPANY**) has established a jobsite inspection program designed to identify safety and health violations and hazards that exist in all the work environments. It is the responsibility of all supervisors to regularly observe jobsites, work areas, tools, and equipment daily, and take all appropriate actions necessary to eliminate or control any hazards that are identified. If eliminating or controlling the hazard goes beyond the authority of the supervisor, the Safety and Health Manager is to be asked for assistance. Both (**COMPANY**) employees and non-company personnel must be kept away from every identified hazard until it is corrected.

The frequency of inspections will be based upon potential hazards and the type and complexity of the project. The frequency determination will be made by the Project Manager, Superintendent or the Safety and Health Manager.

It is important to note that more accidents result from unsafe acts or actions than from unsafe conditions. In addition to inspecting jobsites and work areas for unsafe conditions, supervisors should observe operations, work procedures, and employee actions. Unsafe activities must be eliminated and replaced with safe procedures. ***All (COMPANY) employees committing unsafe acts or violating safe working procedures are to be reprimanded according to the Disciplinary Action Policy of Section 2.4.***

All hazards identified during daily observations or periodic inspections should be corrected or controlled immediately. The report of such inspections may be included in the Project's ***Daily Field Report*** that includes all activities on the jobsite, or may be completed using the ***Jobsite Safety Inspection Report Form*** of Section 4.2.3.

All field reports and inspection reports noting any safety deficiency or safety concern are to be forwarded to the Safety and Health Manager for review and follow-up action when required.

The Safety and Health Manager shall make both scheduled and surprise inspections. The results of his or her inspection will be discussed with the project management and supervisory staff immediately following the inspection. Supervisors will take immediate action to eliminate, correct, or control the hazard, or ensure that project management has taken such action.

4.2.2 Field Inspection Procedures

The manner in which a field inspection is conducted depends upon its purpose and scope. The following is a list of guidelines and suggestions that will help ensure that each field safety inspection is comprehensive and efficient:

- Be aware in advance of what you plan to inspect.
- Review applicable regulations. Familiarize yourself with the hazards that are associated with the operation or equipment that you intend to inspect.
- Schedule the inspection at a time that will allow a maximum opportunity to view operations and work practices. Mid-morning or early afternoon are often good times.
- Be alert to all hazards, and do not merely run through the checklist. A checklist is only a reminder. Hazards unique to a specific situation should not be overlooked.
- Choose a systematic inspection route. Cover the entire area footprint and leave nothing out. When re-inspecting a work area, approach from a different direction or use a different route to gain a different perspective of jobsite conditions.
- Take notes, and be sure to note the exact description and/or location of every hazard when observed. Include ideas for corrective action. Do not wait until after the inspection to record hazardous conditions or unsafe actions; details might get forgotten.
- Look for the source cause(s) of adverse conditions and practices. Think in terms of correction action only. Do not focus on fixing blame.
- If an unsafe piece of equipment or condition is observed, the supervisor shall warn the employees of the hazard(s) involved. If a life-threatening hazard exists, the operation must be immediately suspended. The work will be allowed to resume only after the supervisor is satisfied that the hazard has been thoroughly corrected.

- Any unsafe conditions on the part of a subcontractor must be brought to that company's immediate attention. All such unsafe conditions must be reported in writing and noted in the ***Daily Field Report*** or field ***Safety Inspection Report***. All reports are to be forwarded to the subcontractor, with a copy sent to the (**COMPANY**) Safety and Health Manager. ***Always remember that you and (COMPANY) could be held accountable for the actions and negligence of every subcontractor associated with our work.***

4.2.3 Sample Jobsite Safety Inspection Report Form

The report of all jobsite inspections may be included in the project's ***Daily Field Report***, or may be completed using the ***Sample Jobsite Safety Inspection Report Form*** of this section (sample appears on the following page).

NOTE TO MANUAL PREPARER: Check with your own insurance provider to see if they have or can recommend a specific form to be used for the purposes of this sample section.

COMPANY LOGO	Construction Safety and Loss Control Program Manual	Page: 4-5	Appr:
	4. JOBSITE SAFETY AND HEALTH PROGRAM	Revision Date:	

Sample

(COMPANY) Jobsite Safety Inspection Report Form

Project: ______________________ Date: ______________ S M T W T F S

Project #: ______________________

Location: ______________________

Project Manager: ______________________

Superintendent: ______________________

Observed Hazard: ______________________

____

____

____

____

Responsibility: ☐ (COMPANY) ☐ Subvendor: ______________________

Hazard Location: ______________________

____

____

____

Recommended Corrective Action: ______________________

____

____

____

____

____

____

Work Recommended to be Stopped in Hazard Location: ☐ YES ☐ NO

Work Stopped in Hazard Location: ☐ YES ☐ NO

Corrective Action Completed: Date: ______________ Time: ____________

Inspection performed by: ______________________

cc: ☐ Safety & Health Manager
☐ Project Manager
☐ Superintendent
File: ☐ PM________
☐ CF________

COMPANY LOGO	Construction Safety and Loss Control Program Manual	Page: 4-6	Appr:
	4. JOBSITE SAFETY AND HEALTH PROGRAM	Revision Date:	

4.3 JOBSITE SAFETY PLANNING CHECKLIST

Prior to the startup of every construction project, consideration must be given to the basic logistics of safety in and around the jobsite. As the project develops in the field, the initial determinations and arrangements for each of the items may require periodic or special modification in order to ensure that all safety considerations remain adequate as conditions on the jobsite change. At a minimum, the items listed below must be addressed. Where necessary, action must be taken during all appropriate stages of job development—prior to, during, and after construction.

The following is a checklist of specific items to be considered during the development of the specific jobsite safety program. It is a helpful aid, but is not necessarily conclusive. Check with the Safety and Health Manager, and *review the specific project requirements*.

1. Administration: Establish and maintain on-file adequate supplies of:
   a. Site Safety Program _____
   b. Safety Review Checklists _____
   c. Posted listings of emergency services _____
   d. Accident Investigation Report Forms _____
   e. Accident Eyewitness Statement Outline _____
   f. Tailgate Safety Meeting Agendas _____
   g. OSHA forms 101 & 200 (See Section 8) _____

2. Forward "Job Startup Notice" to **(COMPANY)** insurance carrier _____

3. Identify all formal and available informal safety personnel:
   a. **(COMPANY)** Safety Officer _____
   b. **(COMPANY)** On-Site Safety Representative _____
   c. Owner Safety Representative _____
   d. All **(COMPANY)** employees with first aid, CPR or other safety training _____
   e. All owner employees with first aid, CPR or other safety training _____
   f. All sub-vendor employees with first aid, CPR or other safety training _____

4. Identify safety and emergency services:
   a. Notify of the presence of the jobsite, anticipated work force, and duration of the project:
      1) Fire Department _____
      2) Police Department _____
      3) Medical facilities _____

COMPANY LOGO	Construction Safety and Loss Control Program Manual	Page: 4-7	Appr:
	4. JOBSITE SAFETY AND HEALTH PROGRAM	Revision Date:	

b. Identify locations and emergency travel routes to:
   1) Emergency rooms ______
   2) Non-emergency medical facilities ______
   3) Fire Department ______
   4) Police Department ______

c. Listings of local services:
   1) Physicians ______
   2) Eye experts ______
   3) Orthopedics ______
   4) Paramedics ______

5. Unique owner requirements (in addition to "usual" considerations):
   a. Construction parking ______
   b. Occupied area parking ______
   c. Limited parking or access ______
   d. Strict security ______
   e. "Hot work" permits ______
   f. Full or part-time safety person ______
   g. Insurance wrap-up policy ______
   h. Other: ________________________ ______

6. Traffic control—a plan to address:
   a. Entering/leaving of the construction work force ______
   b. Material/equipment deliveries ______
   c. Equipment access and storage ______
   d. Street and site cleanup and maintenance ______
   e. Adequate warning signs and other postings ______

7. Utility location protection—locate and protect/relocate:
   a. Water ______
   b. Gas ______
   c. Electric ______
   d. Telephone ______
   e. Telex/cable ______
   f. Communication ______
   g. Storm/sanitary sewer ______

8. Preexisting condition survey:

Prior to the start of any construction, record all preexisting damage by whatever means appropriate, including engineered surveys, photographs, or videotape recordings.

COMPANY LOGO	Construction Safety and Loss Control Program Manual	Page: 4-8	Appr:
	4. JOBSITE SAFETY AND HEALTH PROGRAM	Revision Date:	

a. Initial survey should be made by the Site Superintendent.
b. Determine the need for additional detailed inspection(s). _____
c. If sufficient evidence of damage is present, consider retaining a professional photographer to record it. _____
d. If extreme or unique circumstances, determine if a professional engineer is/may be needed to produce surveys or other documentation. _____
e. Prior to proceeding with (c) or (d), secure approval from **(COMPANY)** Senior Project Manager. _____
f. In every case, photograph *and* videotape the entire site, as well as all approach roads and routes. Prior to site mobilization and start of any construction activity, include:
   1) Surrounding buildings _____
   2) Roads _____
   3) Utilities _____

9. Plan the locations and configurations of site services and their relative proximities:
   a. Field offices _____
   b. Material staging and storage areas _____
   c. Fuel storage and fuel distribution arrangements _____
   d. Traffic control _____
   e. Administrative and worker parking _____
   f. Pedestrian access _____
   g. Location and configuration of temporary utilities _____
   h. Ongoing temporary power arrangements _____
   i. Temporary lighting—Site _____
   j. Temporary lighting—Other construction areas _____
   k. Temporary heat _____
   l. Temporary power _____
   m. Welding and cutting torches (friendly fire) _____

10. Identify specific owner requirements that may exceed customary considerations, such as:
    a. Special access requirements or needs _____
    b. Unique security requirements _____
    c. Involvement of owner or other designated safety or security personnel _____
    d. Special insurances or legal considerations _____
    e. Special notices or other required communications _____

11. Determine necessary provisions for protection of the public:
    a. Site fencing _____

b. Lighting ______
c. Signs and notices ______
d. Traffic control ______
e. Guardrails ______
f. Walkways (covered/uncovered) ______

12. Determine all fire protection needs. Project-specific planning should include:
   a. Appropriate ABC-type fire extinguishers:
      1) Sizes ______
      2) Quantities ______
      3) Locations ______
   b. Fire-protected storage cabinets and/or areas for vital, sensitive, or special files or materials ______
   c. Establishment and maintenance of all appropriate fire protection measures in accordance with OSHA requirements for specific work operations ______

13. Consider necessary protection of the site and building (if there is one):
   a. Signs and notices posted ______
   b. Special walkways and traffic provisions ______
   c. Barricades and other safety barriers ______
   d. Fences, guardrails, and canopies ______
   e. Parking, traffic and walkway lighting ______
   f. All provisions for handicap access ______
   g. Security and safety personnel ______

4.4 VENDOR INSURANCES

4.4.1 Coverages

All contractors, subcontractors, trade contractors, and independent individuals performing *any* work on the site must be covered by all insurances to the specified limits as enumerated in the Contract Documents, *as described in their specific agreements for services*, or as otherwise required by (**COMPANY**).

It is the responsibility of the Project Manager and the Site Superintendent to confirm the presence of all such insurances and to secure the correct and complete certificates of such insurances from each entity on the site before allowing *any* work to proceed.

There are *no* exceptions.

4.4.2 Additional Insured

It is a standard requirement of **(COMPANY)** that all subvendor agreements provide that **(COMPANY)** is specifically named as additional insured on the respective policy.

It is the responsibility of the Project Manager—through the Project Engineer if one is assigned to the project—that **(COMPANY)** is actually named as additionally insured. Note that the specific language "additionally insured" is the only acceptable language. Terms such as "certificate holder" or other language not specifically noting "additionally insured" will not be accepted.

NOTE TO MANUAL PREPARER: Check with your own insurance carrier for specific language and documentation required to identify and confirm minimum coverage requirements acceptable to your company. Describe those specifics in this sample section.

4.5 FIRST AID KIT

(COMPANY) will provide a complete medical first aid kit on each jobsite for use by company personnel. All contractors, subcontractors, trade contractors, and individuals (non-**(COMPANY)** employees) on the site are required to provide and maintain their own first aid kits in manner adequate for the type of work being performed. It is the responsibility of the **(COMPANY)** Project Manager, Site Superintendent, and each trade foreman to enforce this requirement on the parts of each subvendor on every jobsite.

4.6 SAFETY AND FIRST AID TRAINING

NOTE TO MANUAL PREPARER: As of this writing, there is a certain minimal requirement for employers to provide safety training for their employees, but not necessarily to the extent described in the sample section that follows. The material is presented as an example of one company's approach, but should only be included in your own actual Safety and Loss Control Program Manual if your company management agrees with the ideas expressed. Otherwise, the material must be modified as appropriate.

4.6.1 Basic First Aid and CPR

All (**COMPANY**) personnel are encouraged to take basic first aid training and to become certified in first aid and CPR (cardio-pulmonary resuscitation). The Superintendent, Project Manager, all trade foremen, and everyone in any supervisory capacity should feel a particular responsibility in this area.

(**COMPANY**) will arrange for in-house company training in basic first aid and CPR from time to time. It is not necessary for any company employee to wait for this convenience, however. Contact the local Red Cross, area hospitals, and local fire departments to identify medical first aid training programs available. Participation by every (**COMPANY**) employee throughout every jobsite and office is strongly encouraged.

4.6.2 Reimbursement for Training Expenses

In most instances, the cost of certain first aid training programs may be reimbursed to the (**COMPANY**) employee if the program of interest had been first coordinated with the (**COMPANY**) Safety and Health Manager, and approved in advance by (**COMPANY**) senior management.

If any (**COMPANY**) employee wishes to arrange for such basic or advanced first aid training on their own, it is the policy of (**COMPANY**) to reimburse the employee 100% of the tuition and direct expenses associated with the training program. The employee need only to present the program certificate or other written documentation that the program has been successfully completed by the employee.

4.7 JOBSITE SAFETY MEETINGS

4.7.1 General

It is the policy of (**COMPANY**) that regular jobsite safety meetings be conducted by the Site Superintendent, trade foreman, and/or the Safety and Health Manager to:

- Orient all (**COMPANY**) and subvendor tradespeople to project requirements and objectives.

- Identify areas on the site that present special problems or concerns, and determine the best corrective action for each situation.
- Swap or trade safety and health topics of both general and specific nature. Develop conversation and discussion, and documentation that will further the safety education and awareness of all project participants.

The **(COMPANY)** Jobsite Safety Meeting program series is arranged in three levels:

1. The Initial Project Safety Orientation Meeting
2. Ongoing **(COMPANY)** Jobsite Safety Meetings
3. Periodic (weekly) Jobsite "Tailgate" Safety Meetings

4.7.2 Initial Project Safety Orientation Meeting

NOTE TO MANUAL PREPARER: The conduct and documentation of a certain minimum amount of safety meetings may be prescribed in special programs offered by insurance carriers. Check with your own insurance provider for specific help or instruction in the final arrangement and configurations of all company safety meetings.

The Project Safety Orientation Meeting is conducted by the Site Superintendent at the very onset of the project. It can be its own stand-alone meeting, or it can be combined with an early Superintendent's Meeting with all **(COMPANY)** and subvendor trade foremen.

The meeting must be given the complete attention by all project participants that it deserves. All subvendors will be required to participate. Attendance will be documented.

At a minimum, items to be reviewed at the meeting include:

1. All safety requirements of the **(COMPANY)** Contract, and the stated requirements of each vendor's contracts and subcontracts.
2. **(COMPANY)**, project, and special rules as they relate to the specific project.
3. The jobsite utilization program, specifically with regard to arrangements for stored materials, proper materials handling, traffic, access, security, communication, etc.
4. Fire protection requirements and procedures; general consideration and specific "hot work" areas.
5. Evacuation procedures; alarms, routes, communication, personnel, etc.

6. Posting of all emergency phone numbers.
7. Designation of each company's Safety Officer and Alternate; phone numbers, contact procedure, etc.
8. Identification of all jobsite personnel with any medical training. Issue of hard hat red crosses.
9. Review of required first aid and medical supplies; inventory of current jobsite supplies and necessary supplements.
10. Specific jobsite precautions with respect to protection of workers and protection of the public; temporary protection, fall protection, hard hat, safety shoes, gloves, eye protection, etc.
11. Jobsite security issues and arrangements; fences, site and field office area lighting, alarms, etc.
12. Notification to all (**COMPANY**) and subvendor participants that willful and/or repeated violations may be grounds for layoffs or ejection from the jobsite.
13. Notification of the expectation of regular participation in periodic safety meetings, and of the need for all trade foremen to periodically conduct their own Tailgate Safety Meetings and to immediately forward minutes of those meetings to (**COMPANY**).

4.7.3 Ongoing Jobsite Safety Meetings

Regular ongoing Jobsite Safety Meetings can be conducted as their own stand-alone arrangements. It may, however, be more effective and more easily accomplished if the meetings are combined with the regular Superintendent's Meetings with all trade foremen as the *first* routine item on each meeting agenda.

Meeting agenda items regarding safety and health will at a minimum include:

1. Review of any observed actual or possible safety violations. Determination and implementation of corrective measures, and responsibility for correction.
2. Any planned or other changes in the jobsite utilization program as previously established.
3. Jobsite housekeeping and cleanup status; protection, safety, and security.
4. Review of stored materials and equipment.
5. Review of past or upcoming hot work, and the handling of burning and welding equipment. Review of hot work permit requirements.
6. Review of temporary power and maintenance, and any necessary corrections or upcoming changes.
7. Review of temporary heat and temporary fuel requirements and handling.
8. Review of all safety performances throughout the project.

Consider combining this meeting with a Tailgate Safety Meeting as described in the following section. You already have everyone's attention; use it.

4.7.4 Weekly Tailgate Safety Meetings

NOTE TO MANUAL PREPARER: The content and format of the Weekly Tailgate Safety Meetings may be prescribed or suggested by government agencies, your insurance carrier, or other safety authority. Take advantage of available assistance to edit the requirements and suggestions of this sample section.

Each Site Superintendent—or Crew Foreman if that is the senior site position—is responsible to conduct *weekly* safety training sessions with all field personal under that individual's authority. Such meetings may be coordinated by the **(COMPANY)** Safety and Health Manager for assistance in such presentations, but the responsibility remains with the senior **(COMPANY)** site representative.

Topics will be relevant and timely with respect to the perceived current needs and all general concerns. Attendance will be recorded by signature, and be kept on-file by the Safety and Health Manager. Persistent absences by any **(COMPANY)** employee will be recorded and the information distributed to that employee's personnel record.

The Tailgate Safety Meeting is a 10 to 15 minute discussion, conducted for the purposes of:

1. Disseminating information included in the regular Safety Meetings.
2. Reviewing jobsite safety and cleanup conditions and establishment of corrective measures.
3. Discussion of specific safety items for the training and education of all tradespeople. When possible, the items should be relevant to some observed job condition and ongoing work practices, but they can be of random nature as well.

4.7.5 (COMPANY) Tailgate Safety Bulletin

As an aid in the organization, presentation, and recording of each Tailgate Safety Meeting, a **(COMPANY)** ***Tailgate Safety Bulletin*** similar to the sample on the following page will be sent periodically to the jobsite by Safety and Health Manager.

Highlights of each bulletin include:

- One topic per bulletin;
- Clear outline of issues and ideas to be discussed;

- Area provided (on the back) to record signatures of each Safety Meeting attendee.

(Sample (**COMPANY**) Tailgate Safety Bulletin)

NOTE TO MANUAL PREPARER:** **Any jobsite safety bulletins can be prepared by the company as a convenience to all company field staff. Each safety bulletin can be an outline of the topics to be covered in Weekly Tailgate Safety Meetings.

Refer to Sample Section 4.7.6 of this Safety and Loss Control Program Manual as follows for a short listing of suggested topics for Weekly Tailgate Safety Meetings. Contact your local OSHA office, workmen's compensation insurance provider, and your industry and trade associations for additional assistance in the preparation of specific lists and topic outlines.

4.7.6 Suggested Tailgate Safety Meeting Topics and Outlines

Whether or not a Tailgate Safety Bulletin is used, the following list is an aid in the organization and development of the jobsite Tailgate Safety Meeting Program.

One topic should be covered per meeting. Suggested outlines can include:

1. Emergency care and procedures to obtain aid. Determination of qualifications for administering first aid.

2. Controlling and restoring breathing. Aiding a choking victim.

3. Aiding a burning victim. Aiding a shocked victim.

4. Aiding a victim with a broken bone.

5. Fire regulations at the work site.
   a. Smoking and nonsmoking areas
   b. Location and use of fire fighting equipment
   c. Periodic check of fire extinguisher "charges"
   d. Storage and use of combustible materials

6. Chains and slings
   a. Care and proper use

7. On-site accidents
   a. Falls—causes and prevention
   b. Jewelry—rings, chains, etc.
   c. How to manually lift loads safely

8. Trench safety
   a. Shoring requirements
   b. Cave-ins
   c. Rescue procedure

9. Compressed gas cylinders
   a. Dangers of compressed air
   b. Handling and storage of cylinders

10. Use of friendly fire
    a. Temporary heat and fuel
    b. Temporary power
    c. Proper use of cutting and welding equipment
    d. Jobsite "hot work" permit requirements

11. Personal safety rules and equipment
    a. Hard hat
    b. Safety shoes
    c. Gloves
    d. Proper clothing
    e. Proper use of safety belts and nets
    f. Goggles and eye injuries

12. Safe handling of power tools
    a. Rip saws
    b. Bench grinders
    c. Drills
    d. Power-actuated tools

13. Heavy equipment
    a. General safe use
    b. General equipment safe operating procedures

14. Cranes and rigging
    a. General safe use

15. Earth-moving equipment

COMPANY LOGO	Construction Safety and Loss Control Program Manual	Page: 4-17	Appr:
	4. JOBSITE SAFETY AND HEALTH PROGRAM	Revision Date:	

   a. General safe use
   b. Operation near embankments
   c. Operation in confined areas

16. Hazardous materials—industrial hygiene
    a. Asbestos—detection and reaction
    b. Treatment and disposal of hazardous or controlled materials
    c. PCP
    d. Carbon monoxide
    e. Lead

17. Confined-area activity
    a. Definitions of confined area
    b. Egress
    c. Ventilation
    d. Evacuation

18. Scaffolding and staging
    a. Erecting procedures
    b. Working on it
    c. Use arrangements

19. Ladders
    a. Types and uses
    b. Areas of use

20. Protection of the public
    a. Walkways
    b. Lighting
    c. Barricades
    d. Signs and notices

21. General emergency safety procedures
    a. General procedures
    b. Accidents involving serious injury or death
    c. Property damage accidents

22. Hazard communication
    a. The (COMPANY) HAZ-COM Manual
    b. Hazardous chemicals on the jobsite
    c. MSDS's
    d. Labels and warning systems

23. Basic first aid topics
    a. Identification of on-site personnel with safety training
    b. Availability of safety training programs.

4.8 JOBSITE SAFETY REVIEW CHECKLIST

The Jobsite Safety Review Checklist that follows is prepared as a convenience to aid the Site Superintendent, the Safety and Health Manager, and the Project Manager in their periodic and routine reviews of each jobsite in order to help identify possible conditions that may cause or contribute to an accident.

It is by no means any guarantee that all possible hazards will be identified or accommodated, nor is it a list that specifically complies with any regulatory requirements, but is a list of those items that are frequently observed and easily identified.

Check OSHA regulations, those of all appropriate authorities having jurisdiction over the jobsite, and, through the **(COMPANY)** home office, with our insurance carrier.

The Checklist may include many items that may not necessarily apply to a particular project, or that may otherwise might apply to projects of a nature other than the specific jobsite being considered each time. Even so, reviewing the entire list periodically may help to identify issues previously overlooked, and that may now apply.

Finally, the use of the Checklist and any other site safety reviews should not be confined strictly to **(COMPANY)** operations:

- For practical purposes, and to aid in the most comprehensive perspective, consider the entire site as if it were in the custody of **(COMPANY)**, whether or not it is legally the case.
- Observe the conditions being provided for by any general contractor, construction manager, other subcontractors, and any separate contractors that may also be operating on the site.
- Report observed violations or possible violations immediately to your **(COMPANY)** supervisor, in order that the entire site may be made more safe for **(COMPANY)**, all other workers, and the public.

COMPANY LOGO	Construction Safety and Loss Control Program Manual	Page: 4-19	Appr:
	4. JOBSITE SAFETY AND HEALTH PROGRAM	Revision Date:	

Jobsite Safety Review Checklist

1. Signs, Notices, & Notifications
- a. Safety signs in place ____
- b. Emergency phone numbers ____
- c. Evacuation plan ____
- d. Warnings & instructions to public ____
- e. Restricted access areas ____
- f. Exits ____
- g. No smoking ____
- h. Electrical dangers ____
- i. Personal protective equipment req'd ____
- j. Operating instructions ____
- k. Flammable materials ____
- l. Hazardous materials ____
- m. Danger areas ____
- n. Trenches ____
- o. All personnel & occupants notified to expect loud noises ____
- p. ____________ ____
- q. ____________ ____

2. Overhead Protection
- a. Entrances ____
- b. Warnings ____
- c. Construction ____
- d. ____________ ____
- e. ____________ ____

3. Hoisting Equipment
- a. Guys ____
- b. Cables & sheaves ____
- c. Turnbuckles ____
- d. Signals ____
- e. Carcover & enclosure ____
- f. Ladder ____
- g. Car arresting device ____
- h. Base barricade ____
- i. Platforms ____
- j. Clear staging areas ____
- k. ____________ ____
- l. ____________ ____

4. Walkways & Ramps
- a. Adequate construction ____
- b. Width ____
- c. Railings ____
- d. Curbs & toe boards ____
- e. Slope & rise limit ____
- f. Non-slip treads & tactile areas ____
- g. ____________ ____
- h. ____________ ____

5. Ladders
- a. Construction ____
- b. Secure placement ____
- c. Cleats ____
- d. Landings ____
- e. Hand-holds ____
- f. Cages ____
- g. ____________ ____

6. Excavations & Trenches
- a. Shoring ____
- b. Slope reposes ____
- c. Ladders ____
- d. Stockpile of excavated material ____
- e. Removal of excavated material ____
- f. Barricades & railings ____
- g. Tunnels ____
- h. Blasting arrangements ____
- i. Excavations properly dewatered ____
- j. Approved shoring designs ____
- k. Proper ventilation; free of toxic fumes ____
- l. ____________ ____
- m. ____________ ____

7. Fire Protection
- a. Storage of flammable materials ____
- b. Container markings ____
- c. Temporary heaters ____
- d. Compressed gas cylinders ____
- e. Tar kettles ____
- f. Welding equipment & operations ____
- g. Fire extinguishers (quantity/type) ____
- h. Fire safety equipment ____
- i. ____________ ____
- j. ____________ ____

8. Openings - Walls, Floors, Roofs ____
- a. Perimeter railing ____
- b. Toe boards ____
- c. Tight covers ____
- d. Flaggings ____
- e. ____________ ____
- f. ____________ ____

9. Scaffolds
- a. Construction ____
- b. Secure placement ____
- c. Railings & toeboards ____
- d. Rigging ____
- e. Safety lines, belts, rope guards ____
- f. ____________ ____
- g. ____________ ____

10. Stairs & Landings
- a. Adequate construction ____
- b. Temporary treads ____
- c. Clear of debris ____
- d. Proper rise/run ____
- f. Railings ____
- g. ____________ ____
- h. ____________ ____

11. Material Handling
- a. Size/bulk ____
- b. No sharp edges ____
- c. Weight limits ____
- d. Team lifting ____
- e. ____________ ____

COMPANY LOGO	Construction Safety and Loss Control Program Manual	Page: 4-20	Appr:
	4. JOBSITE SAFETY AND HEALTH PROGRAM	Revision Date:	

12. Housekeeping
- a. Nails, debris ____
- b. Tool storage & staging ____
- c. Containers ____
- d. Clear aisles & walkways ____
- e. Clean site ____
- f. Dumpster(s) location/condition ____
- g. Proximity of waste storage to hazardous conditions ____
- h. ________________ ____
- i. ________________ ____

13. Lighting & Temporary Wiring
- a. Lighting ____
- b. Wire height ____
- c. Proper grounding ____
- d. Wire connections ____
- e. Overcurrent protection ____
- f. Extension cords in good repair ____
- g. Extension cords & temp. power receptacles using GFI's ____
- h. Temp. power closed to weather ____
- i. ________________ ____
- j. ________________ ____

14. Grounding & Electrical Equipment
- a. Correct grounding ____
- b. Ground Fault Interrupters ____
- c. ________________ ____
- d. ________________ ____

15. Portable & Power Saws
- a. In good conditions ____
- b. Guards ____
- c. Kickback protection ____
- d. Ventilation ____
- e. Safe fuel procedures ____
- f. ________________ ____
- g. ________________ ____

16. Hand Tools
- a. In good condition ____
- b. Insulated / grounded ____
- c. Projectile tools ____
- d. Power-actuated tools ____
- e. Operators trained in proper use ____
- f. ________________ ____
- g. ________________ ____

17. First Aid
- a. Proper kit size & components ____
- b. Kit supply maintained ____
- c. Trained employees ____
- d. Emergency numbers posted ____
- e. Hospital routes known ____
- f. ________________ ____
- g. ________________ ____

18. Traffic Control
- a. Parking ____
- b. Speed control ____
- c. Barricades ____
- d. Separation of haul roads ____
- e. ________________ ____
- f. ________________ ____

19. Personal Protective Equipment
- a. Hard hats ____
- b. Goggles / safety glasses ____
- c. Gloves ____
- d. Respirators ____
- e. Hearing protection ____
- f. Safety shoes ____
- g. No loose clothing ____
- h. All work areas sanitary ____
- i. ________________ ____
- j. ________________ ____

20. Heavy Equipment
- a. Guards ____
- b. Warning bells ____
- c. Fueling arrangements ____
- d. Ground slope (areas of operation) ____
- e. Rough terrain ____
- f. Operator qualifications ____
- g. ________________ ____
- h. ________________ ____

21. Security
- a. Fencing ____
- b. Lighting ____
- c. Alarm systems ____
- d. Monitoring arrangements ____
- e. Guard service ____
- f. Target equipment ____
- g. Secure equipment practices ____
- h. Police notification procedures ____
- i. ________________ ____
- j. ________________ ____

22. Liability
- a. Release forms executed/delivered for all trades using:
  - Hoists ____
  - Elevators ____
  - Scaffolding ____
  - Equipment ____
- b. Arrange for jobsite inspection by insurance carrier ____
- c. ________________ ____
- d. ________________ ____

23. Other
- a: ________________ ____
- b. ________________ ____

4.8 JOBSITE SAFETY REVIEW CHECKLIST *(continued)*

1. Signs, Notices, and Notifications
   a. Safety signs in place ____
   b. Emergency phone numbers ____
   c. Evacuation plan ____
   d. Warnings and instructions to public ____
   e. Restricted access areas ____
   f. Exits ____
   g. No Smoking ____
   h. Electrical dangers ____
   i. Personal protective equipment req'd ____
   j. Operating instructions ____
   k. Flammable materials ____
   l. Hazardous materials ____
   m. Danger areas ____
   n. Trenches ____
   o. All personnel and occupants notified to expect loud noises ____
   p. ____________________ ____
   q. ____________________ ____
2. Overhead Protection
   a. Entrances ____
   b. Warnings ____
   c. Construction ____
   d. ____________________ ____
   e. ____________________ ____
3. Hoisting Equipment
   a. Guys ____
   b. Cables and Sheaves ____
   c. Turnbuckles ____
   d. Signals ____
   e. Carcover and enclosure ____
   f. Ladder ____
   g. Car arresting device ____
   h. Base barricade ____
   i. Platforms ____
   j. Clear staging areas ____
   k. ____________________ ____
   l. ____________________ ____

COMPANY LOGO	Construction Safety and Loss Control Program Manual	Page: 4-22	Appr:
	4. JOBSITE SAFETY AND HEALTH PROGRAM	Revision Date:	

4. Walkways and Ramps
   a. Adequate construction ____
   b. Width ____
   c. Railings ____
   d. Curbs and toe boards ____
   e. Slope and rise limit ____
   f. Non-slip treads and tactile areas ____
   g. ________________ ____
   h. ________________ ____

5. Ladders
   a. Construction ____
   b. Secure placement ____
   c. Cleats ____
   d. Landings ____
   e. Hand-holds ____
   f. Cages ____
   g. ________________ ____

6. Excavations and Trenches
   a. Shoring ____
   b. Slope reposes ____
   c. Ladders ____
   d. Stockpile of excavated material ____
   e. Removal of excavated material ____
   f. Barricades and railings ____
   g. Tunnels ____
   h. Blasting arrangements ____
   i. Excavations properly dewatered ____
   j. Approved shoring designs ____
   k. Proper ventilation; free of toxic fumes ____
   l. ________________ ____
   m. ________________ ____

7. Fire Protection
   a. Storage of flammable materials ____
   b. Container markings ____
   c. Temporary heaters ____
   d. Compressed gas cylinders ____
   e. Tar kettles ____
   f. Welding equipment and operations ____
   g. Fire extinguishers (quantity/type) ____

COMPANY LOGO	Construction Safety and Loss Control Program Manual	Page: 4-23	Appr:
	4. JOBSITE SAFETY AND HEALTH PROGRAM	Revision Date:	

h. Fire safety equipment ____
i. ____________________ ____
j. ____________________ ____

8. Openings—Walls, Floors, Roofs
   a. Perimeter railing ____
   b. Toe boards ____
   c. Tight covers ____
   d. Flaggings ____
   e. ____________________ ____
   f. ____________________ ____

9. Scaffolds
   a. Construction ____
   b. Secure placement ____
   c. Railings and toeboards ____
   d. Rigging ____
   e. Safety lines, belts, rope guards ____
   f. ____________________ ____
   g. ____________________ ____

10. Stairs and Landings
   a. Adequate construction ____
   b. Temporary treads ____
   c. Clear of debris ____
   d. Proper rise/run ____
   e. Railings
   f. ____________________ ____
   g. ____________________ ____

11. Material Handling
   a. Size/bulk ____
   b. No sharp edges ____
   c. Weight limits ____
   d. Team lifting ____
   e. ____________________ ____

12. Housekeeping
   a. Nails, debris ____
   b. Tool storage and staging ____
   c. Containers ____
   d. Clear aisles and walkways ____

COMPANY LOGO	Construction Safety and Loss Control Program Manual	Page: 4-24	Appr:
	4. JOBSITE SAFETY AND HEALTH PROGRAM	Revision Date:	

e. Clean site ____
f. Dumpster(s) location/condition ____
g. Proximity of waste storage to hazardous conditions ____
h. ____________________ ____
i. ____________________ ____

13. Lighting and Temporary Wiring
a. Lighting ____
b. Wire height ____
c. Proper grounding ____
d. Wire connections ____
e. Overcurrent protection ____
f. Extension cords in good repair ____
g. Extension cords and temp.power receptacles using GFI's ____
h. Temp. power closed to weather ____
i. ____________________ ____
j. ____________________ ____

14. Grounding and Electrical Equipment
a. Correct grounding ____
b. Ground Fault Interceptors ____
c. ____________________ ____
d. ____________________ ____

15. Portable and Power Saws
a. In good conditions ____
b. Guards ____
c. Kickback protection ____
d. Ventilation ____
e. Safe fuel procedures ____
f. ____________________ ____
g. ____________________ ____

16. Hand Tools
a. In good condition ____
b. Insulated/grounded ____
c. Projectile tools ____
d. Power-actuated tools ____
e. Operators trained in proper use ____
f. ____________________ ____
g. ____________________ ____

COMPANY LOGO	Construction Safety and Loss Control Program Manual	Page: 4-25	Appr:
	4. JOBSITE SAFETY AND HEALTH PROGRAM	Revision Date:	

17. First Aid
    a. Proper kit size and components ____
    b. Kit supply maintained ____
    c. Trained employees ____
    d. Emergency numbers posted ____
    e. Hospital routes known ____
    f. ____________________ ____
    g. ____________________ ____

18. Traffic Control
    a. Parking ____
    b. Speed control ____
    c. Barricades ____
    d. Separation of haul roads ____
    e. ____________________ ____
    f. ____________________ ____

19. Personal Protective Equipment
    a. Hard hats ____
    b. Goggles/safety glasses ____
    c. Gloves ____
    d. Respirators ____
    e. Hearing protection ____
    f. Safety shoes ____
    g. No loose clothing ____
    h. All work areas sanitary ____
    i. ____________________ ____
    j. ____________________ ____

20. Heavy Equipment
    a. Guards ____
    b. Warning bells ____
    c. Fueling arrangements ____
    d. Ground slope (areas of operation) ____
    e. Rough terrain ____
    f. Operator qualifications ____
    g. ____________________ ____
    h. ____________________ ____

21. Security
    a. Fencing ____
    b. Lighting ____

COMPANY LOGO	Construction Safety and Loss Control Program Manual	Page: 4-26	Appr:
	4. JOBSITE SAFETY AND HEALTH PROGRAM	Revision Date:	

c. Alarm systems ____
d. Monitoring arrangements ____
e. Guard service ____
f. Target equipment ____
g. Secure equipment practices ____
h. Police notification procedures ____
i. ____________________ ____
j. ____________________ ____

22. Liability
a. Release forms executed/delivered for all trades using: ____
Hoists ____
Elevators ____
Scaffolding ____
Equipment ____
b. Arrange for jobsite inspection by insurance carrier ____
c. ____________________ ____
d. ____________________ ____

23. Other
a. ____________________ ____
b. ____________________ ____

4.9 JOBSITE SAFETY INSPECTIONS BY OUTSIDE OFFICIALS

The following procedure is to be used in the event an outside official enters a (**COMPANY**) jobsite with the intent of inspecting or investigating jobsite safety. Such outside officials can include:

- OSHA inspectors
- State inspectors
- Fire inspectors
- Worker's Compensation investigators
- Insurance investigators

Note that the intent of this procedure is not in any way to obstruct or otherwise make any inspection more difficult. The actions discussed in this section are only

intended to allow the company to exercise our legal right to proper representation to certain government inspections, and to be sure that all facts are properly considered.

The (**COMPANY**) procedure upon such a visit is as follows:

1. Direct the inspector to the highest ranking (**COMPANY**) employee present on the jobsite at that time.

2. The highest ranking (**COMPANY**) employee will then:
   a. Ask the inspector of his or her complete intentions, and for proper identification.
   b. Write down the inspector's name, title, organization, ID or badge number, date, and the time of day.
   c. Request the inspector to delay further activity until the (**COMPANY**) Safety and Health Manager or other appropriate (**COMPANY**) official can be dispatched to the jobsite.
   d. Contact the (**COMPANY**) Safety and Health Manager and the (**COMPANY**) Senior Project Manager, and provide the information obtained to that point. If either individual is unavailable for any reason, so notify the other. If both individuals are unavailable for any reason, immediately contact company senior management.
   e. Offer a place for the inspector to sit and wait, if possible. Should the inspector insist on proceeding without the Safety and Health Manager or other (**COMPANY**) official present, note the time and accompany him or her through the entire inspection. Take comprehensive notes of the entire activity.

Above all, every (**COMPANY**) employee is to be courteous, helpful, and fully cooperative with every authorized inspector at all times.

Contact the (**COMPANY**) Safety and Health Manager if there are any questions regarding these situations.

Sample Section 5

FIRST AID, MEDICAL, AND EMERGENCY PROCEDURES

Section Use

The form and content of this sample Section 5 *First Aid, Medical, and Emergency Procedures* is designed as an example of company instruction to its employees with respect to the material presented. It is intended to provide for each employee:

- Description of general company procedures to be used by individuals in an emergency situation
- Information regarding the deployment of jobsite first aid equipment.
- A list of emergency facilities, with telephone contacts and summary directions.
- An easily identifiable listing of emergency and other medical care providers who are participants in any company medical insurance program that may be available to its employees

The manual preparer should secure the list of health care providers from the company human resources department. If feasible, consider arranging the listing in the final manual both alphabetically and by area or region.

Also include the list of all available hospital or other emergency rooms and ambulance services. Be sure that each listing includes directions from the nearest main street and interstate highways.

The sample section that follows is an example of a company safety and loss control program approach to the objectives indicated above. The sample language and forms are to be considered in the context of your own company and its manner of operation.

Carefully consider these objectives during the review and editing process of the section material in order to customize the section content to the specific needs of your organization. Prepare final language that is consistent with decisions made by your own management in the determination of your own company program.

It is important to understand that safety rules and regulations change from time to time. Be sure to review the final draft of the section and the entire program manual with appropriate safety, insurance, and legal professionals prior to issue and use. Periodically review your program with company management and all appropriate professionals to be certain that the company program remains current with all requirements.

COMPANY LOGO	Construction Safety and Loss Control Program Manual	Page: 5-1	Appr:
	5. FIRST AID, MEDICAL, AND EMERGENCY PROCEDURES	Revision Date:	

(Sample)

SECTION 5

First Aid, Medical, and Emergency Procedures

5.1 FIRST AID PROCEDURE

5.1.1 Minor First Aid Treatment

All job related injuries and illnesses, no matter how minor, must be reported to the employee's immediate supervisor. Where minor first aid treatment is required and adequate first aid supplies are available at the jobsite, office, shop, etc., seeking outside medical facility help is not mandatory.

5.1.2 First Aid Kits

Approved first aid kits must be present and adequately supplied at every **(COMPANY)** main office, shop, remote offices, and in most of our company on-road type vehicles at all times. They are available at the **(COMPANY)** main office or by contacting the company Safety and Health Manager.

5.1.3 First Aid Qualified Employees

(COMPANY) has several first aid qualified employees throughout the company. Many of these employees have elected to display a red cross decal on his or her hard hat stating "Certified First Aid." These employees each have a wallet-sized card from the qualifying organization.

On all remote job sites where outside emergency medical assistance is further than (20) minutes away (i.e., walk-in clinic, hospital, doctor's office, etc.), the company shall have at least one approved first aid kit, and endeavor to have at least one first aid qualified employee on the site, if possible.

All **(COMPANY)** employees are urged to participate in certified CPR and/or basic first aid training. All employees are eligible for reimbursement of 100% of the cost of such training. Refer to Section 4.6.2 of this *Safety and Loss Control Program Manual* for additional information regarding such training and reimbursement.

5.1.4 Mobile Communication

Each jobsite with a field office has a permanent telephone installed. It is available for company use only for regular business purposes, but is available for use by anyone during a safety or medical emergency.

For those jobsites that do not have a permanent field office, the senior company employee on the jobsite shall make arrangements for a mobile telephone, and to be sure that the mobile telephone is in close proximity at all times. The mobile telephone is to be used for company business only; or by anyone in the event of a safety or medical emergency.

5.2 EMERGENCY PROCEDURE (INJURY)

The general procedure to follow in the case of an injury that is serious enough to require professional medical attention is as follows:

- Immediately attend to the injured's life-threatening conditions.
- Call 911, or send someone to call 911 if a phone is not available for any reason. Give your exact location, and the specific details of the emergency.

NOTE TO MANUAL PREPARER: Use the section below if your company uses radios in its daily communication between the members of the work force.

- If a phone is not available, use a company radio to call the base station. Give your Unit number, and say "**Unit (number) to base—I have an emergency.**" This will let the base station operator and other units know the urgency of the call. Repeat as needed (all other units are to remain silent, unless assistance is required by you). Give the base station operator your exact location, type of emergency, and what you need. The base station operator or person receiving the radio transmission will remind others as needed to "**keep the radio clear—emergency in progress.**" The base station operator will also give the "**All clear—emergency over**" transmission.
- Be certain that the (**COMPANY**) Safety and Health Manager is notified of the emergency at the earliest opportunity. Secure the accident scene until released by the Safety and Health Manager or (**COMPANY**) company principal.

5.3 MEDICAL ATTENTION

If there is a need to send a (**COMPANY**) employee to a medical facility for a job related injury or illness ***other than an emergency***, we currently use the ***(insert the name of the company worker's compensation insurance and managed care provider)***.

An injured employee must never be allowed to drive or otherwise transport himself or herself to any off-site area or facility. The injured person must be driven by a (**COMPANY**) employee in good physical condition and with a valid driver's license to the nearest approved medical care provider.

Under ***(insert name of state in which the Company operates)*** law, the employer has the right to choose the authorized medical provider in all cases except emergencies. We have found the ***(insert the name of the company worker's compensation insurance and managed care provider)*** to provide excellent service to our employees, and are a great advantage strategically for most of our work areas.

Every employee intending to use outside medical attention should make every effort to contact the (**COMPANY**) Human Resources Department for pre-authorization and direction.

Contact: ***(Individual's Name)***, Human Resources Director
(Individual's Name), Safety and Health Manager

This communication is desired but not essential if not practical or possible. In such cases, refer to Section 5.5 following for the complete listing of approved medical care providers for your own determination of the most appropriate and expedient medical care location.

5.4 HOSPITAL EMERGENCY ROOMS

Emergency care facilities are to be identified and posted as part of the safety portion of the Jobsite Utilization Program. If it becomes necessary to bring any employee to an emergency room, use the list below as a convenience.

NOTE TO MANUAL PREPARER: Provide a complete list of all hospital emergency rooms and any other emergency care facilities throughout the area(s) of company operation. Include directions to each one from main thoroughfares.

It is also helpful to provide two listings: one by area/location and the other alphabetically.

5.5 MEDICAL CARE PROVIDER LISTING AND LOCATION

The complete list of approved medical care providers in the ***(insert the name of the company worker's compensation insurance and managed care provider)*** listing along with their addresses and telephone numbers, is included in its entirety in the pages that follow.

- The first half is the listing of care providers arranged ***by area.***
- The second half is the same listing arranged ***alphabetically.***

NOTE TO MANUAL PREPARER: Insert the complete listing of services and providers as identified and explained in this manual section for medical care providers who participate in any company medical benefit package that may be made available to company employees. Note that it is also helpful to provide two listings: one by area/location and the other alphabetically.

Sample Section 6

Accident Investigation, Reporting, and Records

Section Use

The form and content of this Sample Section 6 *Accident Investigation, Reporting, and Records* is designed as an example of company instruction to its employees with respect to the material presented. It is intended to provide for each employee:

- An explanation of required OSHA reporting requirements in the event of jobsite accidents of varying severity.
- Guidelines for performing effective investigations of an accident, and thoroughly reporting all relevant facts.
- Instruction with respect to company policy regarding vehicle and safety procedures.
- Explanation of company policy regarding disciplinary action and safety award programs

The sample section that follows is an example of a company safety and loss control program approach to the objectives indicated above. The sample language and forms are to be considered in the context of your own company and its manner of operation.

To a great extent, the material presented is the result of the interpretation of OSHA regulations in effect as of the time of the sample manual preparation, and effort to summarize their effects into a set of instructions. It is important that the preparer of any actual safety and loss control program Manual evaluate the material in the context of current information and developments.

Carefully consider these objectives during the review and editing process of the section material in order to customize the section content to the specific needs of your organization. Prepare final language that is consistent with decisions made by your own management in the determination of your own company program.

It is important to understand that safety rules and regulations change from time to time. Be sure to review the final draft of the section and the entire program manual with appropriate safety, insurance, and legal professionals prior to issue and use. Periodically review your program with company management and all appropriate professionals to be certain that the company program remains current with all requirements.

COMPANY LOGO	Construction Safety and Loss Control Program Manual	Page: 6-1	Appr:
	6. ACCIDENT INVESTIGATION, REPORTING, AND RECORDS	Revision Date:	

(Sample)

SECTION 6

Accident Investigation, Reporting, and Records

COMPANY LOGO	Construction Safety and Loss Control Program Manual	Page: 6-2	Appr:
	6. ACCIDENT INVESTIGATION, REPORTING, AND RECORDS	Revision Date:	

6.1 ACCIDENT INVESTIGATION AND REPORTING REQUIREMENTS

All accidents and occupational illnesses must be reported.

It is the policy of (**COMPANY**) that every accident or injury of any kind and at any level of severity *must* be immediately reported by the employee to the Crew Foreman, Site Superintendent, or Project Manager—even if the employee believes the incident to be minor. Each employee must be aware that the failure to report an incident on the day that it occurs may result in the denial of workmen's compensation claims if the situation later worsens. Disciplinary action by the company may become necessary, as well.

In the case of serious injury, such as fractures, cuts requiring sutures, multiple injuries, unconsciousness, vision impairment, fatality, poison ingestion, respiratory distress, etc., the (**COMPANY**) home office and the Safety and Health Manager must be notified as soon as practical—immediately *after* appropriate emergency measures have been taken.

For any injury to any (**COMPANY**) employee while working, a (**COMPANY**) ***Accident Investigation Report*** form must be completed by the appropriate individual(s) and be filed immediately with the (**COMPANY**) Human Resources Department. The Human Resources Department will communicate with the company insurance carrier.

NOTE TO MANUAL PREPARER: Verify current OSHA reporting requirements, and modify this portion of the sample Safety and Loss Control Program Manual as necessary.

If an on-the-job accident occurs that results in the death of an employee or in the hospitalization of three (3) or more employees, all employers on the project site, regardless of the number of employees, are required to report the accident, in detail, to the following:

- To OSHA's toll-free number: **1–800–321–OSHA**, or
- To the nearest OSHA office.

These reports must be made within eight (8) hours of the incident. This requirement applies to each such fatality or hospitalization of three (3) or more employees which occurs within thirty (30) days of an incident. The only exception to this requirement is if the employer does not learn of a reportable incident at the time it occurs and the incident would otherwise be reportable, the employer shall make the report within eight (8) hours of the time the incident is reported to any agent or employee of the employer.

6.1.1 OSHA Forms and Their Use

NOTE TO MANUAL PREPARER: Verify current OSHA reporting requirements, and modify this portion of the sample Safety and Loss Control Program Manual as necessary.

Company safety records are maintained by (**COMPANY**) on a calendar-year basis. They must be maintained for five (5) years at (**COMPANY**) offices, and must be available for inspection by representatives of OSHA or other designated state agency. They are not routinely sent to OSHA or any other government agency.

Two OSHA forms are required to be maintained:

OSHA No. 200 Log and Summary of Occupational Injuries and Illnesses

Each recordable occupational injury and illness must be logged on this form within six (6) working days from the time the employer learns of it. If the log is prepared at a central location, a copy that is current to within forty-five (45) calendar days must be present at all times at (**COMPANY**) offices. A substitute for the OSHA No. 200 form may be acceptable if it contains the OSHA 200 form information at a minimum, and is as detailed and understandable as the OSHA No. 200 form.

It is the responsibility of the Safety and Health Manager to maintain the OSHA 200 log form.

OSHA No. 101 Supplementary Record of Occupational Injuries and Illnesses

The OSHA No. 101 form contains much more detail about each injury or illness. It also must be completed within six (6) working days from the time the employer learns of the work-related injury or illness. A substitute for the OSHA No. 101 form (such as insurance or workers' compensation forms, or the ***(COMPANY) Accident Investigation Report*** of Section 6.4) may be used if it contains all the information included in the OSHA No. 101 form.

Annual Survey

At any time, an employer may be selected to participate in the annual statistical survey. If selected, each employer will receive in the mail the OSHA No. 200S form for this purpose, soon after the close of the year. If selected, (**COMPANY**) will complete the report using the OSHA No. 200 form as the source of information, and process the form as required.

Many specific OSHA standards have additional recording and reporting requirements. Check the appropriate category in this (**COMPANY**) ***Safety and Loss Control Program Manual***, and verify requirements with the (**COMPANY**) Safety and Health Manager.

An ***Accident Investigation Report*** form appears in Section 6.4.

6.1.2 OSHA Posting Requirements

A copy of the totals and information on the last page of the OSHA No. 200 form for the year must be posted at the (**COMPANY**) employee lounge and at every jobsite with an established field office. This copy must be posted no later than February 1, and kept in place until March 1. Even though there were no injuries or illnesses during the year, zero must be entered on the totals line, and the form posted.

It is the responsibility of the Safety and Health Manager to ensure that all such posting requirements are complied with and maintained.

6.1.3 Supplementary Record of Occupational Injuries and Illnesses (OSHA Form 101)

To supplement the ***Log and Summary of Occupational Injuries and Illnesses (OSHA No. 200)***, (**COMPANY**) must maintain a record of each recordable occupational injury or illness. Workers' compensation, insurance, and other reports such as the (**COMPANY**) ***Accident Investigation Report*** form of Section 6.4 are acceptable as records if they contain all facts listed below or are supplemented to do so.

If no report in appropriate form is made for other purposes, the ***Supplementary Record of Occupational Injuries and Illnesses (OSHA No. 101)*** may be used to record necessary information or the necessary acts can be listed on a separate plain sheet of paper.

These records must be maintained by the Safety and Health Manager at the (**COMPANY**) home office, and must be available for examination by representatives of the Department of Labor and the Department of Health, Education, and Welfare. The records must be maintained for a minimum period of five (5) years following the end of the calendar year to which they apply.

Such records must contain the minimum information:

1. ***Information regarding the employer***
   Name.
   Mailing address.
   Location if different from mailing address.

2. ***Information regarding the injured or ill employee***
   Name.
   Social Security number.
   Home address.
   Age.
   Sex.
   Occupation.
   Department.

3. ***Information regarding the occupational injury or illness***
   Description of injury or illness, including part of body affected.
   Name of the object or substance that directly injured the employee.
   Date of injury or diagnosis of illness.

4. ***Other information***
   Name and address of physician.
   If hospitalized, name and address of hospital.
   Date of the report.
   Name of the person preparing the report.

See definitions on the back of the OSHA No. 200 form for additional information.

The OSHA Form 101 on the following page is to be used to comply with the requirements of this section as described above.

COMPANY LOGO	Construction Safety and Loss Control Program Manual	Page: 6-6	Appr:
	6. ACCIDENT INVESTIGATION, REPORTING, AND RECORDS	Revision Date:	

OSHA No. 101
Case or File No.

Form approved
OMB No. 44R 1453

Supplementary Record of Occupational Injuries and Illnesses

EMPLOYER

1. Name: ____________________
2. Mail Address: ____________________
3. Location, if different from mail address: ____________________

INJURED OR ILL EMPLOYEE

4. Name: ____________________ Social security No.: ____________
5. Home Address: ____________________
6. Age: __________ 7. Sex (Check One): Male _____ Female _____
8. Occupation: ____________________
   (Enter regular job title, not the specific activity he or she was performing at the time of injury)
9. Department: ____________________
   (Enter the name of department or division in which the injured person is regularly employed, even though her or she may have been temporarily working in another department at the time of injury)

THE ACCIDENT OR EXPOSURE TO OCCUPATIONAL ILLNESS

10. Place of accident or exposure: ____________________
    (No. and Street) (City or Town) (State)
    If accident or exposure occurred on employer's premises, give address of plant or establishment in which it occurred. Do not indicate department or division within the plant or establishment. If accident occurred outside employer's premises at an identifiable address, give that address. If it occurred on a public highway or at any other place which cannot be identified by number and street, please provide place references locating the place of injury as accurately as possible.
11. Was place of accident or exposure on employer's premises? ____________ (Yes or No)
12 What was the employee doing when injured? (Be specific. If he was using tools or equipment or handling material, name them and tell what he was doing with them.) ____________________
13. How did the accident occur? (Describe full the event which resulted in the injury or occupational illness. Tell what happened and how it happened. Name any objects or substances involved and tell how they were involved. Give full details on all factors which led or contributed to the accident. Use separate sheet of paper for additional space.) ____________________

OCCUPATIONAL INJURY OR OCCUPATIONAL ILLNESS

14. Describe the injury or illness in detail and indicate the part of body affected. (e.g.: amputation or right index finger at second joint; fracture of ribs; lead poisoning; dermatitis of left hand; etc.) ____________________
15. Name the object or substance which directly injured the employee: (For example, the machine or thing he struck against or which struck him; the vapor or poison he inhaled or swallowed; the chemical or radiation which irritated his skin; hernias, etc., the thing he was lifting, pulling, etc.) ____________________
16. Date of injury or initial diagnosis of occupational illness: ____________________
17. Did employee die? __________ (Yes or No)

OTHER

18. Name and address of physician: ____________________
19. If hospitalized, name and address of hospital: ____________________

Date of Report: ____________________

Prepared by: ____________________
Official Position: ____________________

6.1.4 Sample OSHA Fatality/Multiple Injury Hospital Notification Report

The ***OSHA Fatality/Multiple Injury Hospital Notification Report*** displayed on the following page is to be used in order to comply with the reporting requirements described in this Section 6.

This form must be prepared by the (**COMPANY**) supervisory or management personnel most familiar with the facts of the incident, and be telephoned in to OSHA ***by the (COMPANY) Safety and Health Manager*** within the following time frames:

- Within (8) hours after the death of any employee from a work related incident, or
- Within (8) hours after the in-patient hospitalization of (3) or more employees as a result of a work related incident.

NOTE: This requirement applies to each such fatality or hospitalization of (3) or more employees which occurs within thirty (30) days of an incident. An *exception, however, will be cases in which the **(COMPANY)** does not learn of a reportable incident at the time it occurred; such report shall be made within (8) hours of the time the incident is reported to any agent or employee of (**COMPANY**).*

COMPANY LOGO	Construction Safety and Loss Control Program Manual	Page: 6-8	Appr:
	6. ACCIDENT INVESTIGATION, REPORTING, AND RECORDS	Revision Date:	

OSHA Fatality/Multiple Injury Hospital Notification Report

CALL 1-800-321-OSHA
OR CONTACT THE LOCAL OSHA AREA OFFICE

Date: ______________________ Time of Accident: __________ AM PM

Employer: ☐ (COMPANY) ☐
____

Location of Accident: ______________________________
____

Number of: Fatalities: ________ and/or Hospitalized Employees: ________

(COMPANY) Contact Person: ______________________ Phone #:
____

OSHA Contact Person: ______________________________

Brief Description of Accident: ______________________________
____
____
____
____
____

Prepared By: Sign: ______________________

6.1.5 OSHA Form 200: Log and Summary of Occupational Injuries and Illnesses

The OSHA Form 200: Log and Summary of Occupational Injuries and Illnesses, on the following pages are to be used to comply with the requirements of Section 6.1.1. A supply of Form 200 can be secured from OSHA by calling 1/800–321–OSHA.

(Place OSHA Form 200: Log and Summary of Occupational Injuries and Illnesses here)

COMPANY LOGO	Construction Safety and Loss Control Program Manual	Page: 6-10	Appr:
	6. ACCIDENT INVESTIGATION, REPORTING, AND RECORDS	Revision Date:	

Bureau of Labor Statistics
Log and Summary of Occupational
Injuries and Illnesses

NOTE: **This form is required by Public Law 91-596 and must be kept in the establishment for *5 years*. Failure to maintain and post can result in the issuance of citations and assessment of penalties. *(See posting requirements on the other side of form.)***

RECORDABLE CASES: You are required to record information about every occupational **death**, every nonfatal occupational **illness**, and those nonfatal occupational **injuries** which involve one or more of the following: loss of consciousness, restriction of work or motion, transfer to another job, or medical treatment (other than first aid). ***(See definitions on the other side of form.)***

Case or File Number	Date of Injury or Onset of Illness	Employee's Name	Occupation	Department	Description of Injury or Illness
Enter a nonduplicating number which will facilitate comparisons with supplementary records.	Enter Mo./day.	Enter first name or initial, middle initial, last name.	Enter regular job title, not activity employee was performing when injured or at onset of illness. In the absence of a formal title, enter a brief description of the employee's duties.	Enter department in which the employee is regularly employed or a description of normal workplace to which employee is assigned, even though temporarily working in another department at the time of injury or illness.	Enter a brief description of the injury or illness and indicate the part or parts of body affected. Typical entries for this column might be: Amputation of 1st joint right forefinger; Strain of lower back; Contact dermatitis on both hands; Electrocution—body.
(A)	(B)	(C)	(D)	(E)	(F)
					PREVIOUS PAGE TOTALS
					TOTALS (Instructions on other side of form.)

SPECIMEN FORM

OSHA No. 200

FOLD

COMPANY LOGO	Construction Safety and Loss Control Program Manual	Page: 6-11	Appr:
	6. ACCIDENT INVESTIGATION, REPORTING, AND RECORDS	Revision Date:	

U.S. Department of Labor

For Calendar Year 19 ______ Page ____ of ____

Company Name

Establishment Name

Establishment Address

Form Approved
O.M.B. No. 1220-0029

Extent of and Outcome of INJURY						Type, Extent of, and Outcome of ILLNESS												
Fatalities	Nonfatal Injuries					Type of Illness							Fatalities	Nonfatal Illnesses				
Injury Related	Injuries With Lost Workdays				Injuries Without Lost Workdays	CHECK Only One Column for Each Illness *(See other side of form for terminations or permanent transfers.)*							Illness Related	Illnesses With Lost Workdays				Illnesses Without Lost Workdays
Enter **DATE** of death. Mo./day/yr.	Enter a CHECK if injury involves days away from work, or days of restricted work activity or both.	Enter a **CHECK** if injury involves days away from work.	Enter number of **DAYS** *away from work.*	Enter number of **DAYS** of *restricted work activity.*	Enter a **CHECK** if no entry was made in columns 1 or 2 but the injury is recordable as defined above.	Occupational skin diseases or disorders	Dust diseases of the lungs	Respiratory conditions due to toxic agents	Poisoning (systemic effects of toxic materials)	Disorders due to physical agents	Disorders associated with repeated trauma	All other occupational illnesses	Enter **DATE** of death. Mo./day/yr.	Enter a **CHECK** if illness involves days away from work, or days of restricted work activity, or both.	Enter a **CHECK** if illness involves days away from work.	Enter number of **DAYS** *away from work.*	Enter number of **DAYS** of *restricted work activity.*	Enter a **CHECK** if no entry was made in columns 8 or 9.
(1)	(2)	(3)	(4)	(5)	(6)	(7) (a)	(b)	(c)	(d)	(e)	(f)	(g)	(8)	(9)	(10)	(11)	(12)	(13)

INJURIES SPECIMEN FORM ILLNESSES

Certification of Annual Summary Totals By ______ Title ______ Date ______

FOLD

OSHA No. 200

POST ONLY THIS PORTION OF THE LAST PAGE NO LATER THAN FEBRUARY 1.

COMPANY LOGO	Construction Safety and Loss Control Program Manual	Page: 6-12	Appr:
	6. ACCIDENT INVESTIGATION, REPORTING, AND RECORDS	Revision Date:	

Instructions for OSHA No. 200

I. Log and Summary of Occupational Injuries and Illnesses

Each employer who is subject to the recordkeeping requirements of the Occupational Safety and Health Act of 1970 must maintain for each establishment a log of all recordable occupational injuries and illnesses. This form (OSHA No. 200) may be used for that purpose. A substitute for the OSHA No. 200 is acceptable if it is as detailed, easily readable, and understandable as the OSHA No. 200.

Enter each recordable case on the log within six (6) workdays after learning of its occurrence. Although other records must be maintained at the establishment to which they refer, it is possible to prepare and maintain the log at another location, using data processing equipment if desired. If the log is prepared elsewhere, a copy updated to within 45 calendar days must be present at all times in the establishment.

Logs must be maintained and retained for five (5) years following the end of the calendar year to which they relate. Logs must be available (normally at the establishment) for inspection and copying by representatives of the Department of Labor, or the Department of Health and Human Services, or States accorded jurisdiction under the Act. Access to the log is also provided to employees, former employees and their representatives.

II. Changes in Extent of or Outcome of Injury or Illness

If, during the 5-year period the log must be retained, there is a change in an extent and outcome of an injury or illness which affects entries in columns 1, 2, 6, 8, 9, or 13, the first entry should be lined out and a new entry made. For example, if an injured employee at first required only medical treatment but later lost workdays away from work, the check in column 6 should be lined out, and checks entered in columns 2 and 3 and the number of lost workdays entered in column 4.

In another example, if an employee with an occupational illness lost workdays, returned to work, and then died of the illness, any entries in columns 9 through 12 should be lined out and the date of death entered in column 8.

The entire entry for an injury or illness should be lined out if later found to be nonrecordable. For example, an injury which is later determined not to be work related, or which was initially thought to involve medical treatment but later was determined to have involved only first aid.

III. Posting Requirements

A copy of the totals and information following the fold line of the last page for the year must be posted at each establishment in the place or places where notices to employees are customarily posted. This copy must be posted no later than ***February 1 and must remain in place until March 1.***

Even though there were no injuries or illnesses during the year, zeros must be entered on the totals line, and the form posted.

The person responsible for the ***annual summary totals*** shall certify that the totals are true and complete by signing at the bottom of the form.

IV. Instructions for Completing Log and Summary of Occupational Injuries and Illnesses

Column A – CASE OR FILE NUMBER. Self-explanatory.

Column B – DATE OF INJURY OR ONSET OF ILLNESS.
For occupational injuries, enter the date of the work accident which resulted in injury. For occupational illnesses, enter the date of initial diagnosis of illness, or, if absence from work occurred before diagnosis, enter the first day of the absence attributable to the illness which was later diagnosed or recognized.

Columns C through F – Self-explanatory.

Columns 1 and 8 – INJURY OR ILLNESS-RELATED DEATHS.
Self-explanatory.

Columns 2 and 9 – INJURIES OR ILLNESSES WITH LOST WORKDAYS.
Self-explanatory.

Any injury which involves days away from work, or days of restricted work activity, or both must be recorded since it always involves one or more of the criteria for recordability.

Columns 3 and 10 – INJURIES OR ILLNESSES INVOLVING DAYS AWAY FROM WORK. Self-explanatory.

Columns 4 and 11 – LOST WORKDAYS—DAYS AWAY FROM WORK.
Enter the number of workdays (consecutive or not) on which the employee would have worked but could not because of occupational injury or illness. The number of lost workdays should not include the day of injury or onset of illness or any days on which the employee would not have worked even though able to work.

NOTE: For employees not having a regularly scheduled shift, such as certain truck drivers, construction workers, farm labor, casual labor, part-time employees, etc., it may be necessary to estimate the number of lost workdays. Estimates of lost workdays shall be based on prior work history of the employee AND days worked by employees, not ill or injured, working in the department and/or occupation of the ill or injured employee.

Columns 5 and 12 – LOST WORKDAYS—DAYS OF RESTRICTED WORK ACTIVITY.
Enter the number of workdays (consecutive or not) on which because of injury or illness:

(1) the employee was assigned to another job on a temporary basis, or
(2) the employee worked at a permanent job less than full time, or
(3) the employee worked at a permanently assigned job but could not perform all duties normally connected with it.

The number of lost workdays should not include the day of injury or onset of illness or any days on which the employee would not have worked even though able to work.

☆ U. S. GOVERNMENT PRINTING OFFICE : 1987 O 181-504 (64351)

COMPANY LOGO	Construction Safety and Loss Control Program Manual	Page: 6-13	Appr:
	6. ACCIDENT INVESTIGATION, REPORTING, AND RECORDS	Revision Date:	

Columns 6 and 13 – INJURIES OR ILLNESSES WITHOUT LOST WORKDAYS. Self-explanatory.

Columns 7a through 7g – TYPE OF ILLNESS.
Enter a check in only *one* column for each illness.

TERMINATION OR PERMANENT TRANSFER—Place an asterisk to the right of the entry in columns 7a through 7g (type of illness) which represented a termination of employment or permanent transfer.

V. **Totals**

Add number of entries in columns 1 and 8.
Add number of checks in columns 2, 3, 6, 7, 9, 10, and 13.
Add number of days in columns 4, 5, 11, and 12.
Yearly totals for each column (1-13) are required for posting. Running or page totals may be generated at the discretion of the employer.

If an employee's loss of workdays is continuing at the time the totals are summarized, estimate the number of future workdays the employee will lose and add that estimate to the workdays already lost and include this figure in the annual totals. No further entries are to be made with respect to such cases in the next year's log.

VI. **Definitions**

OCCUPATIONAL INJURY is any injury such as a cut, fracture, sprain, amputation, etc., which results from a work accident or from an exposure involving a single incident in the work environment.
NOTE: Conditions resulting from animal bites, such as insect or snake bites or from one-time exposure to chemicals, are considered to be injuries.

OCCUPATIONAL ILLNESS of an employee is any abnormal condition or disorder, other than one resulting from an occupational injury, caused by exposure to environmental factors associated with employment. It includes acute and chronic illnesses or diseases which may be caused by inhalation, absorption, ingestion, or direct contact.

The following listing gives the categories of occupational illnesses and disorders that will be utilized for the purpose of classifying recordable illnesses. For purposes of information, examples of each category are given. These are typical examples, however, and are not to be considered the complete listing of the types of illnesses and disorders that are to be counted under each category.

7a. **Occupational Skin Diseases or Disorders**
Examples: Contact dermatitis, eczema, or rash caused by primary irritants and sensitizers or poisonous plants; oil acne; chrome ulcers; chemical burns or inflammations; etc.

7b. **Dust Diseases of the Lungs (Pneumoconioses)**
Examples: Silicosis, asbestosis and other asbestos-related diseases, coal worker's pneumoconiosis, byssinosis, siderosis, and other pneumoconioses.

7c. **Respiratory Conditions Due to Toxic Agents**
Examples: Pneumonitis, pharyngitis, rhinitis or acute congestion due to chemicals, dusts, gases, or fumes; farmer's lung; etc.

7d. **Poisoning (Systemic Effect of Toxic Materials)**
Examples: Poisoning by lead, mercury, cadmium, arsenic, or other metals; poisoning by carbon monoxide, hydrogen sulfide, or other gases; poisoning by benzol, carbon tetrachloride, or other organic solvents; poisoning by insecticide sprays such as parathion, lead arsenate; poisoning by other chemicals such as formaldehyde, plastics, and resins; etc.

7e. **Disorders Due to Physical Agents (Other than Toxic Materials)**
Examples: Heatstroke, sunstroke, heat exhaustion, and other effects of environmental heat; freezing, frostbite, and effects of exposure to low temperatures; caisson disease; effects of ionizing radiation (isotopes, X-rays, radium); effects of nonionizing radiation (welding flash, ultraviolet rays, microwaves, sunburn); etc.

7f. **Disorders Associated With Repeated Trauma**
Examples: Noise-induced hearing loss; synovitis, tenosynovitis and bursitis; Raynaud's phenomena; and other conditions due to repeated motion, vibration, or pressure.

7g. **All Other Occupational Illnesses**
Examples: Anthrax, brucellosis, infectious hepatitis, malignant and benign tumors, food poisoning, histoplasmosis, coccidioidomycosis, etc.

MEDICAL TREATMENT includes treatment (other than first aid) administered by a physician or by registered professional personnel under the standing orders of a physician. Medical treatment does NOT include first-aid treatment (one-time treatment and subsequent observation of minor scratches, cuts, burns, splinters, and so forth, which do not ordinarily require medical care) even though provided by a physician or registered professional personnel.

ESTABLISHMENT: A single physical location where business is conducted or where services or industrial operations are performed (for example a factory, mill, store, hotel, restaurant, movie theater, farm, ranch, bank, sales office, warehouse, or central administrative office). Where distinctly separate activities are performed at a single physical location, such as construction activities operated from the same physical location as a lumber yard, each activity shall be treated as a separate establishment.

For firms engaged in activities which may be physically dispersed, such as agriculture; construction; transportation; communications; and electric, gas, and sanitary services, records may be maintained at a place to which employees report each day.

Records for personnel who do not primarily report or work at a single establishment, such as traveling salesmen, technicians, engineers, etc., shall be maintained at the location from which they are paid or the base from which personnel operate to carry out their activities.

WORK ENVIRONMENT is comprised of the physical location, equipment, materials processed or used, and the kinds of operations performed in the course of an employee's work, whether on or off the employer's premises.

6.2 ACCIDENT INVESTIGATION REQUIREMENTS

All serious accidents must be investigated. These include all serious accidents to all individuals employed by **(COMPANY)**, the employees of any subcontractor under the control of **(COMPANY)** on the jobsite, and every incident involving any amount of property damage.

This section provides general information and guidelines to be used by the **(COMPANY)** Safety and Health Manager and/or supervisory and management personnel who are closest to the incident.

Immediately after all appropriate emergency measures, first aid, and damage containment measures have been taken, every effort must be made to immediately:

1. Preserve physical evidence.
2. Take photographs and/or videos, and secure other evidence as appropriate.
3. Take statements from the accident victim(s), and eyewitnesses, and anyone who may have any knowledge of definite or possible cause(s) of the accident.

If the accident is serious enough that the **(COMPANY)** insurance carrier will be investigating, assist the investigation in any way requested.

6.3 ACCIDENT INVESTIGATION PROCEDURE

The investigation procedure begins as soon as all immediate danger to people and property has been brought under control, and all **(COMPANY)** and appropriate authorities have been properly notified of the incident. The investigation is to be conducted by the Safety and Health Manager, Project Manager, or their designee.

The purpose of the investigation procedure is to secure and confirm as many specific facts as possible. It is not to place blame. Once the procedure is completed, the causes and conditions can be thoroughly analyzed and determined at a later date.

Each investigator(s) should realize at the onset of each investigation that the information secured at the scene may or may not be complete or accurate for any number of reasons. These reasons can range from sincere difficulties in recollection by witnesses, or even in perception of the event, to deliberate attempts to change, conceal, or omit information. For these reasons, it is essential in good investigating procedure to:

- Keep asking who, what, where, how, why, and when.
- Insert repeated questions of the same item to the same individual for key

considerations. Phrase the question in different ways and spread the repeated questions among other questions and conversation.

- Repeat the same questions to as many different individuals as possible.
- Interview each person in a separate location away from other people; keep independent answers genuinely independent. Eliminate the possibility of two or more individuals consciously or unconsciously "coordinating" their versions of the incident.

The Accident Investigation Procedure is as follows:

1. Identify all individuals who were definitely or may possibly have been in the vicinity of the accident immediately prior to or during the accident.
2. Catalog the names of those individuals, prioritized in the following order:
   a. Those people who were involved in the accident.
   b. Those people who may have caused or contributed to the accident.
   c. Eyewitnesses.
   d. Those people who were in or around the proximity of the accident.
3. Photograph the complete accident area. In addition, videotape if at all possible.
4. Immediately summarize your own understanding of the accident, if any. Include a diagram with as much relevant information as possible. Include appropriate distances, dimensions, and other significant factual information.
5. Use the ***Accident Investigation Report*** form of Section 6.4 of this *Safety and Loss Control Program Manual* as the baseline for the investigation. Complete it after statements have been secured from all people involved.
6. Use the ***Eyewitness Statement Outline*** form of Section 6.5 of this *Safety and Loss Control Program Manual* for each of the individuals identified in items (1) or (2), above.
7. Upon completion of all investigation activities, turn all accident investigation reports, photos, videos, and information over to the Project Manager or the **(COMPANY)** Safety and Health Manager for delivery to appropriate company personnel.

6.4 ACCIDENT INVESTIGATION REPORT

The ***Accident Investigation Report*** form on the following page is to be used as described in the previous section. It is to be completed as thoroughly as possible, and as soon after the incident as practical.

Contact the **(COMPANY)** Safety and Health Manager for assistance, if there are any questions.

COMPANY LOGO	Construction Safety and Loss Control Program Manual	Page: 6-16	Appr:
	6. ACCIDENT INVESTIGATION, REPORTING, AND RECORDS	Revision Date:	

NOTE TO MANUAL PREPARER: The sample Accident Investigation Report provided on the following page is an example of a form that addresses the issues of this sample section. It is important, however, that an appropriate Accident Investigation Form be secured from the company's insurance provider. From there, it may be advisable to use that form exactly as provided, or it may be acceptable to prepare a customized form if desired for any reason—with* more *information on the form than the one provided by the insurance carrier, but not less. If a custom form is prepared, be sure to have it approved by the company insurance carrier prior to adoption.

(Insert the Accident Investigation Report provided by your own insurance provider, or form approved by your insurance provider. The sample form that follows is for illustration purposes only.)

COMPANY LOGO	Construction Safety and Loss Control Program Manual	Page: 6-17	Appr:
	6. ACCIDENT INVESTIGATION, REPORTING, AND RECORDS	Revision Date:	

Sample Accident Investigation Report

(Page 1 of 2)

EMPLOYER: ______________________________

Report Date: ______________ Accident Location: ☐ Home Office ☐ Branch Office ☐ Jobsite ☐ Other

Incident Date: ______________ Address: ______________________________

Incident Time: ______________ AM PM ☐ Vehicle Accident: ☐ Personal Vehicle ☐ Company Vehicle

EMPLOYEE: Name: ______________ Title: ______________ # Years: ______

Home Address: ______________________________

Home Telephone: ______________ Age: ______ Sex: ___M ___F SS#: ______-______-______

USE ADDITIONAL PAGES AS NECESSARY FOR COMPLETE INFORMATION:

1. Witnesses:

Name	Address:	Phone:

2. Complete Description of Accident (Attach separate drawing(s) and/or diagram(s) as appropriate):

3. Individual(s) injured, other than EMPLOYEE named above:

4. Description of injury to EMPLOYEE:

5. Property & equipment damaged in the accident, including the nature of the damage:

6. Describe all personal protective equipment used, and all necessary safety equipment that was in place:

7. Were there any perceived actual or possible safety violations in the area of the accident?:

COMPANY LOGO	Construction Safety and Loss Control Program Manual	Page: 6-18	Appr:
	6. ACCIDENT INVESTIGATION, REPORTING, AND RECORDS	Revision Date:	

Sample Accident Investigation Report

(Page 2 of 2)

8. Describe events and conditions immediately before the accident:

9. Describe events and conditions that did or may have caused or contributed to the accident:

10. Are these conditions still in existence:

11. Were any unsafe actions performed by the EMPLOYEE or by others, or were any safety rules violated in a manner that caused or contributed to the accident?

12. Describe steps taken to secure the area, or otherwise prevent similar occurrence:

13. Describe measures to be taken to prevent reoccurrence:

14. Medical care was provided:
    - ☐ On-site of the accident. By: ____________
    - ☐ Emergency Room: ____________
    - ☐ Hospital: ____________
    - ☐ Doctor's Office: ____________

    Describe care provided:

15. Did EMPLOYEE report to work the next scheduled work day: ☐ YES ☐ NO

16. Other relevant information:

Prepared By:

Name ____________ Title: ____________ Date: ____________

6.5 EYEWITNESS STATEMENT OUTLINE FORM

The sample ***Eyewitness Statement Outline Form*** is to be used by (**COMPANY**) project management and supervisory staff as an aid to assist each eyewitness in providing statements that are complete, organized, and focused. It is in no way to be used to guide or force any respective statement, or as any means to add, delete, or modify any information. To the contrary, all eyewitnesses must be repeatedly encouraged to put any descriptions strictly into their *own* words.

The Eyewitness Statement Outline Form includes categories for all minimum information requirements, and should therefore not be considered to be all-inclusive. Each eyewitness must be encouraged to provide *all* relevant information beyond the minimums outlined in the form.

Contact the (**COMPANY**) Safety and Health Manager for assistance, if you have any question.

NOTE TO MANUAL PREPARER: The Sample Eyewitness Statement Outline Form provided on the following page is an example of a form that addresses the issues of this sample section. It is important, however, that an appropriate Eyewitness Statement Outline Form be secured from the company's insurance provider, if any such form is available. From there, it may be acceptable to prepare a customized form if desired for any reason—with **more** ***information on the form than the one provided by the insurance carrier, but not less. If a custom form is prepared, be sure to have it approved by the company insurance carrier prior to adoption.***

COMPANY LOGO	Construction Safety and Loss Control Program Manual	Page: 6-20	Appr:
	6. ACCIDENT INVESTIGATION, REPORTING, AND RECORDS	Revision Date:	

Eyewitness Statement Outline

Name: ______________________________
Residence: ____
Home Phone: ________________________ Age: ______ Soc. Sec. #: __________________

Employed By: ________________________ Position: ___________________ # Years: ____
Employer's Address: ____

My relationships and acquaintances with any of the parties involved in the accident are as follows:
____
____

The accident happened on (date) ______________ at (time) ______________ AM PM
The weather was (clear, foggy, rainy, etc.) ________________________________
The road was (sandy, wet, dry, potholed, etc.) _____________________________
Immediately before the accident:
- I saw ____
- I heard ____
- I did/was doing ____

During the accident:
- I saw ____
- I heard ____
- I did/was doing ____

Immediately after the accident:
- I saw ____
- I heard ____
- I did/was doing ____

I have examined the diagram attached to this statement, and have shown my position(s), the position(s) of all parties involved, and all relevant details to the best of my knowledge.
Other persons who might have witnessed or may have knowledge relating to the accident are:
____
I (did) (did not) make a statement to the police, opposing counsel, insurance or private investigator, etc.):
____ (If yes, obtain copy)
Supplementary information:
____
____
____
____
____
____
____
____
____
____
____
____
____
____
____
____
____
____

I have read these _______ pages that make up my complete statement. I have made all remarks voluntarily, to the best of my knowledge, and believe them to be true.

Signed: ______________________________________ Date: ____________________

6.6 RETURN TO WORK FROM INJURY OR ILLNESS

It is most important that the (**COMPANY**) main office maintain the ability at all times to clear any employee for a return to work after any job related medical treatment and/or absence for non-job-related hospitalization.

A signed release form from the treating physician is required before any company employee will be allowed to return to work after any job-related medical treatment and/or absence due to non-job-related hospitalization.

No employee may be allowed to return to work unless the (**COMPANY**) supervisor has first been contacted or you have been contacted by the (**COMPANY**) main office with required approval.

If an employee brings a return to work form directly to the employee's supervisor, the form must be brought to the (**COMPANY**) main office for approval. In order to expedite this process, a telephone call to the main office explaining the situation may be acceptable until the form can be brought to the main office. A fax of the form can be helpful, and may satisfy the requirement, subject to the approval of the company Safety and Health Manager or Human Resource Manager.

No company manager or supervisor is to allow any employee to return to work on his or her own word that he or she has a physician's release, or that he or she had provided the (**COMPANY**) main office with such a release. Always contact the main office to verify.

Immediately contact the (**COMPANY**) Safety and Health Manager or Human Resource Manager if there are any questions.

6.7 NEAR-MISS INCIDENTS AND SAFETY CONCERNS

If *any incident* occurs on any company jobsite that could have resulted in a serious job-related injury but did not, such as an excavation cave-in, gas can explosion, serious fall, vehicle accident, equipment rollover, etc., and so on, *the Safety and Health Manager must be notified* as soon as possible, and in any event before the end of that day.

Similarly, if there are any perceived safety issues, any job-related safety concerns, or if any (**COMPANY**) employee wishes to report any known or suspected safety or health hazard, the employee should immediately notify his or her immediate supervisor and/or the company Safety and Health Manager.

6.8 EQUIPMENT AND VEHICLE DAMAGE

Very often, in the case of minor damage to vehicles or equipment where there are no injuries, near-miss incidents or safety concerns will be the result (refer to Section 6.7 of this *Safety & Loss Control Program Manual*). In such cases, only the immediate supervisor needs to be contacted by the observing employee, or in his or her absence, the company Safety and Heath Manager.

Exceptions would be major damage to (**COMPANY**) equipment, to the property of others, or any injury—however slight that damage or injury may appear to be. These occasions require the (**COMPANY**) Safety and Health Manager to be notified immediately, and a complete accident investigation performed in accordance with the requirements of Section 6 of this ***Safety and Loss Control Program Manual.***

6.9 DISCIPLINARY GUIDELINES

NOTE TO MANUAL PREPARER: The disciplinary guidelines presented in this sample section are examples of an approach that may be taken by a company to deal with a very important and sometimes difficult issue. The ideas must be considered very carefully, and reviewed in detail with all appropriate labor, human resources, insurance, and legal professionals before any policy and procedure is finalized in this regard.

The following guidelines are to effect a consistent and recognized disciplinary policy to be implemented when company rules and safety regulations are willfully violated by any (**COMPANY**) employee.

The guidelines will be used in conjunction with the ***Employee Warning Notice*** of the **Section 2.4** and the following **Section 6.10**.

It is important for all (**COMPANY**) employees to be aware that the purpose of such disciplinary action and company policy with respect to it is only to preserve the most safe and healthy employment environments possible for all (**COMPANY**) employees in all work locations. It is not otherwise to cause or contribute to any undue hardship on the part of the employee who is the subject of such disciplinary action.

The Company Procedure for disciplinary action is detailed in Section 2.4 and summarized below:

1st Warning: Verbal warning of the violation or company concern. The employee's immediate supervisor will be notified. Record of the verbal warning will be placed in the employee's personnel record.

2nd Warning: Written Employee Warning of the violation. Written notice will be issued to the employee, with a copy to employee's supervisor, and to the employee's personnel file. Company policy provides that this warning may be removed from an employee's personnel file after a period of (6) months, if no similar warnings are issued to the employee within that time period.

3rd Warning: Documented (3) day suspension from work without pay, which will be enforced through the employee's immediate supervisor. Record of the incident will be placed in the employee's personnel file.

Additional Warning: Unlimited suspension or termination. Placed into effect upon coordination by the employee's immediate supervisor with the Human Resources Department.

These guidelines are in no way meant to limit or structure disciplinary action in cases of flagrant, abusive, insubordinate, or other extreme cases where swift and decisive action is needed to correct, eliminate, or prevent a situation that may endanger the safety, health, or welfare of other company employees, the employees working in or in the vicinity of company jobsites, or the public.

6.10 EMPLOYEE WARNING NOTICE

Use the sample Employee Warning Notice below as an aid in the documentation described in the Section 6.9, above. A supply of these notices are available from the (**COMPANY**) Human Resources manager and from the Safety and Health Manager.

COMPANY LOGO	Construction Safety and Loss Control Program Manual	Page: 6-24	Appr:
	6. ACCIDENT INVESTIGATION, REPORTING, AND RECORDS	Revision Date:	

6.10 Employee Warning Notice

Use the sample Employee Warning Notice below as an aid in the documentation described in the Section 6.9, above. A supply of these notices are available from the (COMPANY) Human Resources manager and from the Safety & Health Manager.

Employee Warning

Date of Offense: ______________________

NAME: __________________________________ #: ____________ S.S. #: _____________

It is necessary to warn you of the following:

___ Absence	___ Using abusive language	___ Safety Violation
___ Tardiness	___ Breaking company or project rules	___ Carelessness
___ Horseplay	___ Leaving work without permission	___ Abuse of equipment
___ Smoking	___ Refusal to obey direction	___ Other:
___ Fighting	___ Violation of Safety Rules	___________________
___ Loafing	___ Poor work due to carelessness	___________________

Explanation of offense: ____

____

____

____

____

Please be sure that this doesn't happen again. Repetition of the offense or any other offense may result in further disciplinary action, without further notice, which may include employee suspension or discharge.

_____________________________ _____________________________

Employee Signature
Print:

Supervisor Signature
Print:

Date: ______________________________

cc: Safety & Health Manager
Employee Personnel File

6.11 (COMPANY) QUARTERLY JOBSITE SAFETY AWARD

NOTE TO MANUAL PREPARER: Consider the discussion in this section very carefully, specifically in the context of your own intentions. There does not appear to be any legal, technical, or insurance reason to compel any company's subscription to the principle implied here, but the material is presented to illustrate an example of one kind of program that is designed to keep the safety objective in the forefront of all company employees' awareness, and to elevate the safety issue to the place of priority that it deserves.

Each quarter, (**COMPANY**) will express its appreciation for the outstanding effort by a company employee in the area of safety and loss control who, through the nomination of his or her supervisor and in consultation with the other supervisors, has consistently demonstrated initiative, competence, and performance in furthering (**COMPANY**)'s goal of zero accidents on every jobsite and in every office.

That employee will be selected at the Quarterly Supervisor Safety Meeting as described in Section 7.1. The following week, the honored employee will be presented with a Certificate of Appreciation highlighting his or her safety-related performance, along with check in the amount of (**INSERT VALUE**).

The record of the honor will be placed in that employee's permanent file.

6.12 (COMPANY) ANNUAL JOBSITE SAFETY AWARD

NOTE TO MANUAL PREPARER: Consider the discussion in this section very carefully, specifically in the context of your own intentions. There does not appear to be any legal, technical, or insurance reason to compel any company's subscription to the principle implied here, but the material is presented to illustrate an example of one kind of program that is designed to keep the safety objective in the forefront of all company employees' awareness, and to elevate the safety issue to the place of priority that it deserves.

On one occasion each year, at the Annual (**COMPANY**) Employee Safety Meeting as described in Section 7.2, one (**COMPANY**) employee will be recognized for his or her exceptional effort in the area of safety and loss control.

A primary criteria for the award will be consistency of quality attention given safety and loss control throughout the year by the employee.

It is not necessary for an employee to have been a recipient of a Monthly Safety

Award to be eligible for the Annual Award. All **(COMPANY)** employees are eligible.

The potential recipients will be nominated by the Site Superintendents and Project Managers. The final decision will be determined by vote at the Annual **(COMPANY)** Employee Safety Meeting.

The recipient of the award will receive a Certificate of Award, along with a check in the amount of (**$INSERT VALUE**). Record of the award will be placed in the employee's permanent file, and the employee will be honored with his or her name being placed permanently on the plaque displayed at the (**COMPANY**) home office.

Sample Section 7

Ongoing Company Safety Meetings

Section Use

The form and content of the Sample Section 7 *Ongoing Company Safety Meetings* is designed as an example of company instruction to its employees with respect to the material presented. It is intended to provide for each employee:

- The determination of company policy with respect to regular safety meetings.
- Identification of meeting schedules, purposes, and participants.
- Conduct expected of all participants and company employees.

The sample section that follows is an example of a company safety and loss control program approach to the objectives indicated above. The sample language and forms are to be considered in the context of your own company and its manner of operation.

Carefully consider these objectives during the review and editing process of the section material in order to customize the section content to the specific needs of your organization. Prepare final language that is consistent with decisions made by your own management in the determination of your own company program.

It is important to understand that safety rules and regulations change from time to time. Be sure to review the final draft of the section and the entire

program manual with appropriate safety, insurance, and legal professionals prior to issue and use. Periodically review your program with company management and all appropriate professionals to be certain that the company program remains current with all requirements.

COMPANY LOGO	Construction Safety and Loss Control Program Manual	Page: 7-1	Appr:
	7. ONGOING COMPANY SAFETY MEETINGS	Revision Date:	

(Sample)

SECTION 7

Ongoing Company Safety Meetings

COMPANY LOGO	Construction Safety and Loss Control Program Manual	Page: 7-2	Appr:
	7. ONGOING COMPANY SAFETY MEETINGS	Revision Date:	

7.1 QUARTERLY SUPERVISOR SAFETY MEETING

NOTE TO MANUAL PREPARER: Consider the discussion in this section very carefully, specifically in the context of your own intentions. There does not appear to be any legal, technical, or insurance reason to compel any company's subscription to the principle implied here, but the material is presented to illustrate an example of one kind of program that is designed to keep the safety objective in the forefront of all company employees' awareness, and to elevate the safety issue to the place of priority that it deserves.

One day per quarter, all **(COMPANY)** Project Managers, Site Superintendents, and Crew Foremen are required to attend the Quarterly Supervisor Safety Meeting.

The date and time of each meeting will be confirmed by **(COMPANY)** as the time approaches, and the meetings will be held at the **(COMPANY)** Training Center. All affected employees will be given as much notice as possible.

The agenda for each meeting may include the following:

1. Attendance
2. Report of the **(COMPANY)** safety record, and current company-wide safety program:
   a. For the previous quarter
   b. For the year-to-date
3. Discussion of current safety issues affecting ongoing **(COMPANY)** projects
4. Review of any near-miss incidents
   a. How they occurred
   b. Possible effects
   c. What is being done to prevent such issues in the future
5. Nominations and vote for the **(COMPANY)** Quarterly Jobsite Safety Award
6. Suggestions for improvement to the **(COMPANY)** safety and loss control program and communication of its information.
7. Training issue(s) of the quarter
   a. Technical information on relevant jobsite safety issues
8. Suggested Tailgate Meeting topics and materials
9. Consideration of any relevant safety and loss control issue affecting or potentially affecting the company and/or its employees.

Every effort must be made by each affected employee to attend these meetings regularly. Absences will be tolerated for sufficiently excusable reasons, but consecutive absences or numerous absences cannot be tolerated. Continuing problems with absences will be recorded in the respective employee's file.

COMPANY LOGO	Construction Safety and Loss Control Program Manual	Page: 7-3	Appr:
	7. ONGOING COMPANY SAFETY MEETINGS	Revision Date:	

7.2 ANNUAL (COMPANY) EMPLOYEE SAFETY MEETING

During the last quarter of each year, the Annual (**COMPANY**) Employee Safety Meeting is arranged. The date, time, location, and arrangements will be announced to all employees with as much notice as possible.

All Project Managers, Site Superintendents, Crew Foremen, shop employees, and on-site employees are expected to make every possible effort to attend. Attendance will be taken, and the record of absence will be placed in the respective employee's personnel file.

The agenda for each meeting may include:

1. Attendance
2. Report of the (**COMPANY**) safety record—company-wide for the year.
3. Review of the most significant problems and successes throughout the year.
4. Nominations and vote for the (**COMPANY**) Annual Jobsite Safety Award.
5. Suggestions for improvement to the (**COMPANY**) Safety and Loss Control Program.
6. Training issues of particular relevance.
7. Consideration of any relevant safety and loss control issue affecting or potentially affecting the company and/or its employees.

Sample Section 8

The Occupational Safety and Health Act (OSHA)

Section Use

The form and content of Sample Section 8 *The Occupational Safety and Health Act (OSHA)* is designed as an example of company instruction to its employees with respect to the material presented. It is intended to provide for each employee:

- An overview of the Occupational Safety and Health Act, and its relevance to the construction industry.
- General instruction regarding the OSHA inspection procedures and the company's role in them.
- Review of the conduct required and/or expected by each company employee during and after an OSHA jobsite inspection.

The sample section that follows is an example of a company safety and loss control program approach to the objectives indicated above. The sample language and forms are to be considered in the context of your own company and its manner of operation.

Carefully consider these objectives during the review and editing process of the section material in order to customize the section content to the specific needs of your organization. Prepare final language that is consistent with decisions made by your own management in the determination of your own company program.

It is important to recognize the possibility of changes in OSHA regulations, policies, or procedures. It is critical that the manual preparer confirm all remarks of this section and all other sections of this *Safety and Loss Control Program Manual* made with regard to such information and direction that have been presented as an example of how to communicate this kind of information to company employees.

It is important to understand that safety rules and regulations change from time to time. Be sure to review the final draft of the section and the entire program manual with appropriate safety, insurance, and legal professionals prior to issue and use. Periodically review your program with company management and all appropriate professionals to be certain that the company program remains current with all requirements.

COMPANY LOGO	Construction Safety and Loss Control Program Manual	Page: 8-1	Appr:
	8. THE OCCUPATIONAL SAFETY AND HEALTH ACT (OSHA)	Revision Date:	

(Sample)

SECTION 8

The Occupational Safety and Health Act (OSHA)

NOTE: Accident Investigation Reporting is included in Section 6 of this *Safety and Loss Control Manual*

8.1 THE OCCUPATIONAL SAFETY AND HEALTH ACT (OSHA)

The Occupational Safety and Health Act of 1970 was initiated to create the standards of conduct that would improve the safety, health, and welfare of all employees in all industries in the United States. The Occupational Safety and Health Administration is the bureau of the U.S. Department of Labor created to administer the Occupational Safety and Health Act of 1970. Both the Act and the Bureau are commonly referred to as "OSHA."

This federal law was enacted to promote safe and healthful work environments for all employees throughout the United States. The regulations under which the construction industry falls are those standards contained in Section 29 of the Code of Federal Regulations Part 1926. Copies of these standards and other OSHA information are available at any time from the company Safety and Health Manager, and from the local OSHA regional office.

(**COMPANY**) is fully committed to complying with both the stated requirements and the ideals of OSHA in every respect. The (**COMPANY**) Safety and Health Manager will review ongoing developments in OSHA and in improvements in methods to communicate OSHA information, and bring such developments to the attention of company management and employees as appropriate from time to time.

8.2 THE OSHA ON-SITE CONSULTATION PROGRAM

OSHA consultation programs are available to all employers who would like to request assistance in recognizing and correcting safety and health hazards in their workplaces. They are designed to encourage employers to tap into the OSHA skills and expertise, and apply these resources toward directly improving the safety conditions of all construction jobsites.

The on-site consultation program is separate from the OSHA inspection and enforcement effort. No citations are issued or penalties imposed for any safety or health problems observed during the consultation effort. The service is confidential. Employee names, the firm's name, and any information provided about the workplace, in addition to any unsafe or unhealthy working conditions that the consultant uncovers, will not be reported routinely to the OSHA inspection staff. The purpose for this relationship is to encourage employers to participate in the program in the best interest of the company and its employees—without fear of reprisal.

The company's obligation to OSHA under the On-Site Consultation Program is a commitment to correcting serious job safety and health hazards in a timely manner. Confirmation of this commitment will be requested by OSHA prior to any actual site visit.

Note the following:

The OSHA On-Site Consultant Should:

- Assist in the identification of hazards in the workplace.
- Suggest general approaches or options for resolving safety or health issues.
- Assist the company in developing or improving an effective safety policy and health program.
- Identify the kinds of help available from OSHA if further assistance is desired.
- Provide the company with training and education at the workplace, or with available sources of such training.
- Provide the company with a written report summarizing findings.
- Under specified circumstances, recommend (COMPANY) for recognition by OSHA and a one-year exclusion from general schedule enforcement inspections.
- Maintain lists of local providers of safety and industrial hygiene services, and make those lists available to the company upon request.

The OSHA On-Site Consultant Should NOT:

- Issue citations or propose penalties for violations of OSHA standards.
- Report possible violations to OSHA enforcement staff, unless an imminent danger exists and there has been a refusal by the company (through its authorized representative) to correct it.
- Guarantee that the workplace will "pass" any OSHA inspection.

For more information on current OSHA consultation programs and other sources of help, contact:

The Regional Administrator
Occupational Safety and Health Administration
U.S. Department of Labor, Region *(Insert #)*

at: ***(Insert address of OSHA Regional Office)***

Phone: (000) 000–0000

COMPANY LOGO	Construction Safety and Loss Control Program Manual	Page: 8-4	Appr:
	8. THE OCCUPATIONAL SAFETY AND HEALTH ACT (OSHA)	Revision Date:	

8.3 OSHA INSPECTION TYPES

There are two basic types of OSHA inspections:

1. General Inspections

In the general, scheduled inspection, a company is randomly selected by computer. An OSHA inspector then schedules a routine inspection of one or more of that company's jobsites. The inspector is required to comply with the stated procedures for jobsite entry. Once the OSHA inspector's entry to the site is obtained by either permission or warrant, the inspector may then move about that site in an unrestricted manner.

If the OSHA inspector desires to enter any restricted area which may contain hazardous materials or trade secrets, he or she is to be advised to postpone that portion of the inspection until appropriate clearance is obtained from company management. In such an event, the company jobsite representative is to request the OSHA inspector to wait, and immediately contact the company Safety and Health Manager for further direction.

2. Complaint Inspections

A complaint inspection is one in which OSHA has been notified of an observed violation. The OSHA inspector addresses the jobsite or area to target the complaint area. An OSHA inspector who intends to conduct a complaint inspection does not need to obtain a warrant. He or she does, however, need to deliver to the company that is the subject of the inspection a properly completed and executed copy of the OSHA complaint form.

Fatal injuries and complaint inspections are given priority within OSHA.

Individuals who initiate complaint inspections are not required to be named on the complaint form, and may remain anonymous.

The complaint inspection does *not* grant the OSHA inspector free movement within the site. The named company has the right to determine the route to the complaint area, so long as it is not unreasonable.

In the complaint inspection, the OSHA inspector may not expand the investigation into other areas, nor may he or she give attention to non-serious conditions that may be observed along the route taken to the complaint area. It is important to note, however, that if the inspector observes a serious condition along the route, the inspector may involve himself or herself as he or she may determine to be appropriate. (Note that for these purposes, a "serious" condition is one that is likely to result in death or physical damage to any employee or the public.)

All accidents involving disaster, such as death or multiple injuries, fall within the scope of a complaint inspection.

8.4 OSHA RIGHT OF ENTRY TO A JOBSITE

It is the policy of (**COMPANY**) to allow lawfully delegated OSHA or other government inspectors onto company jobsites or into company work areas for the purposes of performing legally proper inspections or investigations. Such inspectors will be required to present appropriate credentials, properly request entry in order to conduct a site inspection of any company project, and agree to allow proper company representation.

This policy is not intended in any way to affect or alter any rights of our subcontractors, trade contractors, or other subvendors. Those separate contractors have the right to request a warrant prior to the inspection of their work areas.

Each (**COMPANY**) subcontractor or trade contractor is to be asked of their intentions to require a warrant from the OSHA representative or inspector prior to any inspection. If any separate contractor does require a warrant, the senior company representative on the jobsite must inform the OSHA inspector prior to allowing such an inspection to commence.

Federal compliance officers, state inspectors, similar personnel, or other individuals are *not* to be harassed or intimidated in any way. Any problems or difficulties that may arise during any on-site inspection that cannot be resolved in a reasonable manner are to be immediately referred to the (**COMPANY**) Safety and Health Manager and/or (**COMPANY**) senior management.

Federal and state safety agencies have the ability to impose severe penalties against individuals and/or companies who do not abide by this policy. Such penalties may include monetary fines and jail terms.

8.5 RIGHTS AND RESPONSIBILITIES

The rights and responsibilities of the employer, the employee, and the OSHA inspector are outlined in this section. They are general descriptions only, and should not be considered to be comprehensive. Refer all related questions to the (**COMPANY**) Safety and Health Manager for additional information.

8.5.1 Employer Responsibilities

Generally, OSHA requires *employers* to be responsible for the following:

a. Comply with standards and all rules adopted by the Department of Labor.
b. Provide and maintain areas of employment that are free of hazards which may cause illness, injury, or death to employees.
c. Provide for all personal protective equipment as needed for general and specific situations.
d. Provide minimum required safety training for employees on a regular basis.
e. Periodically conduct their own Jobsite Safety Inspections.
f. Post notices that advise and inform all employees of their rights as defined by the Occupational Safety and Health Act.
g. Prepare and preserve all appropriate records concerning injuries, accidents, citations, and inspections.

8.5.2 Employee Responsibilities

Generally, OSHA requires *employees* to be responsible for the following:

a. Be familiar with, understand, and comply in every respect with those regulations which pertain to their activities.
b. Promptly report all job-related injuries or accidents to the appropriate company official.
c. Prepare and/or participate in the preparation of all reporting and documentation related to observed safety and health violations or suspected violations.
d. Attend and participate in all scheduled safety meetings to the best of their ability.

8.5.3 Employer Rights

The employer has the right of proper representation during any OSHA jobsite or area inspection. The employer's representative is described in greater detail in this section of the *Safety and Loss Control Program Manual*. He or she may question the actions and statements of the OSHA inspector, and may request clarifications where the actions of the OSHA inspector appear to be contrary to the rules of inspection or be otherwise questionable.

Necessary clarifications and other questions or concerns should be immediately brought to the attention of the **(COMPANY)** Safety and Health Manager, preferably while the OSHA inspector is still on location. The Safety and Health Manager may then direct questions to the OSHA inspector or to the OSHA inspector's office, subject to review with (**COMPANY**) senior management.

8.5.4 Employee Rights

If members of a union, employees have the right of representation through their union during an OSHA inspection or subsequent OSHA investigation. Whether union or not, an employee may answer questions regarding the inspection without concern for punitive actions by the employer.

8.5.5 OSHA Inspector Rights

The OSHA inspector may take photographs, material samples, and measurements as they relate to the specific inspection or investigation. He or she may also privately interview employees for the purposes of enforcing the law. The OSHA inspector may not, however, interrupt, suspend, or otherwise disrupt the work, unless a situation of imminent danger is apparent.

8.6 OSHA DOCUMENTATION

The company Safety and Health Manager must maintain all OSHA records and documentation as may be required, including logs of injuries and illness which are job related.

The minimum record forms required by OSHA are the following:

8.6.1 Log of Occupational Injuries and Illnesses (OSHA Form 200)

OSHA requires the *Log of Occupational Injuries and Illnesses* (OSHA Form 200 on page 84) to be established, and that all entries must be current within six working days from the time the employer is notified of an injury or illness occurring on a jobsite, or any portion thereof, that is under the employer's control.

The log does not have to be maintained on any jobsite, but its location must be one that allows the six-day posting requirement to be complied with. A copy of the OSHA Form 200, current to the previous February 28th, must be available for review by the OSHA inspector on each jobsite.

Details of each entry made on OSHA Form 200 must be provided in the OSHA Form 101 as described in Section 8.6.2 below.

Refer to Section 6 *Accident Investigation, Reporting, and Records* of this *Safety and Loss Control Program Manual* for additional information.

8.6.2 Supplementary Record of Occupational Injuries or Illnesses (OSHA Form 101)

The *Supplementary Record of Occupational Injuries or Illnesses* (OSHA Form 101 on page 80) is a detailed record of job related individual injuries or illness. It is required to be completed for each entry recorded on OSHA Form 200 as described in section 8.6.1 above.

NOTE TO MANUAL PREPARER: Be sure that the following remark applies to your company before including in your own actual Safety and Loss Control Program Manual.

First Report of Injury forms, provided by the workman's compensation division of our insurance company, may be used if they provide the same information required by OSHA Form 101 at a minimum.

These forms must be maintained on site for projects with an on-site field office. For projects without field offices, the forms must be maintained by the Safety and Health Manager in the manner described for Form 200.

Refer to Section 6 *Accident Investigation, Reporting and Records* of this *Safety and Loss Control Program* Manual for additional information.

8.7 EMPLOYEE ACTION DURING AN OSHA JOBSITE INSPECTION

8.7.1 Introduction

The purpose of this section is to provide information for **(COMPANY)** employees that may help clarify appropriate behavior on the part of each employee in the event of an inspection of a **(COMPANY)** jobsite by an Occupational Safety and Health Administration Compliance Officer. The section information includes instruction in performing the documentation necessary to record information in a manner that will facilitate research at a later date. Such information may eventually become needed documentation to defend against improper accusation of any impropriety or violation.

The forms in this section are to be provided to each (**COMPANY** sufficient quantities to allow for field use during every OSHA inspect graphs and other supplemental documentation are to be attached to the forms when they are delivered to the company Safety and Health Manager or home office as described below.

It is extremely important that everyone preparing any information on these forms recognize that all factual statements must be clear, all information must be complete and correct, that all photographs and videos must be clear, and that measurements must be accurate. Besides the obvious reasons for accuracy, this effort is important because of the fact that there is always a possibility that the information may be introduced into evidence under oath at a formal procedure at a later date.

8.7.2 Jobsite Safety Inspections by OSHA

The following procedure is to be followed by all company employees in the event of an OSHA official entering a (**COMPANY**) jobsite with the intention of inspecting or investigating jobsite safety:

1. Direct the inspector to the highest ranking (**COMPANY**) employee on the jobsite at that time.
2. The highest ranking (**COMPANY**) employee will then:
   a. Ask the inspector of his or her complete intentions, and for proper identification.
   b. Write down the inspector's name, title, ID or badge number, if any, and the time of day.
   c. **Ask the inspector to delay further activity until the (COMPANY) Safety and Health Manager or other appropriate (COMPANY) official can be dispatched to the jobsite.**
   d. Contact the (**COMPANY**) Safety and Health Manager or the (**COMPANY**) Senior Project Manager (in the case of unavailability of the Safety and Health Manager), and provide them with the information obtained to that point.
   e. Offer a place for the inspector to sit and wait, if possible.
   f. Should the inspector insist on proceeding without the (**COMPANY**) Safety and Health Manager or other (**COMPANY**) official present, note the time and accompany him or her through the entire inspection. Take comprehensive notes of the entire activity.

It is important to note that the intent is not to obstruct or otherwise make an inspection more difficult in any way. The procedure as described is necessary only to preserve the company's legal right to proper representation in certain OSHA and other government inspections.

Please be sure to be courteous and cooperative at all times.

Should any situation or condition arise at any time for which appropriate action by you is not clear, immediately contact the **(COMPANY)** Safety and Health Manager or Senior Project Manager for guidance.

8.7.3 Responsibilities of the OSHA Compliance Officer (CO)

The purpose of jobsite inspections by the OSHA Compliance Officer (CO) is to identify conditions and/or activities on the project which he or she considers to be unsafe and in violation of the Construction Safety Regulations, and measure and photograph all relevant physical circumstances.

In order to accomplish those objectives, the Compliance Officer has free access to any location where he or she wishes to go on the project. As many samples, photographs, and measurements can be taken as he or she believes necessary to further the goals of the inspection. The OSHA Compliance Officer can also request from the company copies of any literature, documents, or contracts which relate to safety or health that the Compliance Officer believes relevant to the inspection effort.

The OSHA Compliance Officer may not violate any known safety regulation, and is therefore responsible for providing and wearing personal safety equipment wherever such equipment is required by OSHA. Any failure by the Compliance Officer to comply with the associated OSHA regulations or with the company project safety program as provided to the Compliance Officer may be adequate grounds for not allowing the OSHA Compliance Officer to proceed with an inspection, or for halting an inspection that is already in progress.

Such a circumstance is unusual, but is very serious. If such an incident should occur, it must be considered carefully, and the following procedure should be implemented immediately by the senior company employee on-site:

1. Confirm the case; this is a very serious matter that requires verification before taking further action.
2. Notify the company Safety and Health Manager. Describe the complete set of facts and circumstances. The Safety and Health Manager will make the final determination of following procedure.
3. Subject to confirmation by the Safety and Health Manager, advise the OSHA Compliance Officer that he or she is in violation, and request that he or she wait for the arrival of the company Safety and Health Manager, and that he or she comply with the company project safety program as it has been provided to the Compliance Officer.
4. If the Compliance Officer does not wait as requested, or otherwise proceeds while in apparent violation, advise the Compliance Officer that you consider the inspection to be over. Notify the OSHA Area Director, the **(COMPANY)**

Safety and Health Manager, the (**COMPANY**) Project Manager, and (**COMPANY**) senior management of the occurrence.

The OSHA Compliance Officer may discuss matters of safety and health with employees, but should limit such discussion to information relevant to a complete inspection. Such discussions or related activities, however, should not be allowed to unreasonably disrupt ongoing project activities.

8.7.4 Responsibilities of the (COMPANY) Site Staff

Generally, the Project Manager will serve as the (**COMPANY**) Employer's Representative during an OSHA inspection. In the absence of the Project Manager, the Site Superintendent will serve as the (**COMPANY**) Employer's Representative. In the absence of a full time Site Superintendent position, the senior company employee on the project is to contact the company Safety and Health Manager, who will then designate a employee to serve as the Employer's Representative, or serve as the (**COMPANY**) Representative himself or herself. If the (**COMPANY**) Safety and Health Manager is on the site during any inspection, his or her presence does not reduce or otherwise alter the responsibility of the Project Manager or Site Superintendent if they have been confirmed as the (**COMPANY**) Employer's Representative.

The (**COMPANY**) Representative is to advise (**COMPANY**) senior management and the (**COMPANY**) Safety and Health Manager when an OSHA inspection is scheduled or is beginning, provide written record of the inspection, take photographs and videotape recordings, stay with the inspection until it has been declared to be complete, and keep the Safety and Health Manager currently informed of all developments.

The (**COMPANY**) Representative shall direct all (**COMPANY**) employees who are not required to participate in the investigation to keep away from the inspection activities, and maintain their regular operations. Information given to the OSHA Compliance Officers should be limited to the information requested by them.

The (**COMPANY**) representative should not agree with any statement alleging any violation. If a citation is issued the (**COMPANY**) representative is to "receive" the citation, but is to be clear that the citation is not being "accepted." The Safety and Health Manager is to be immediately notified if any citation is delivered.

8.7.5 Photographs and Videotape Recordings

Photographs are to be taken of every item or area that any OSHA Compliance Officer inspects, as well as the surrounding area. A minimum of two photographs should be taken of each situation. The photographs should include:

- One view taken from the position of the OSHA Compliance Officer, and
- One view taken as an overall picture of the area surrounding the alleged violation.

Take as many photographs as necessary to identify the subject and the complete circumstance.

As an example of the possible effect of photographs, some conditions may appear upon close view to be safety violations. A photograph of the overall area, however, may clarify that the item is not, for example, readily accessible; the close-up photograph by itself can therefore create the appearance of a safety violation where no violation actually exists.

NOTE TO MANUAL PREPARER: Determine if the company will issue cameras for the purposes described in this sample section, and if those cameras will have the automatic date capability as discussed. Confirm or correct the text of this sample section as appropriate.

All company issued cameras have the capability to automatically date-stamp each photograph. This function should be used wherever it is available. In addition, each photograph should have the following minimum information on the back:

1. Project name and job number.
2. City and state.
3. Jobsite location—by *specific* reference (column lines, clear plan indicators, etc.).
4. Date and time of photo.
5. Brief description of the alleged violation.
6. Name of the person taking the photo.
7. Chronological number of the photo.

The (**COMPANY**) representative must maintain a complete log of each photograph with this information as the photographs are being taken, in order to ensure the accuracy of the information when it is transposed to the developed photographs. One method of accomplishing this efficiently is to record the photograph information directly on adhesive labels that have been numbered to coincide with the photograph numbers. The labels need only to be transferred to the backs of the photographs immediately after they have been developed.

Whenever possible, supplement the photographs with a simultaneous videotape recording of the inspection. Audiotape recording equipment has historically been disallowed by OSHA, and so the video (with audio) procedure may be similarly objected to by the OSHA Compliance Officer.

If videotape recording is not objected to by the OSHA Compliance Officer, at a minimum, video:

- The route taken by the CO.
- The inspection procedure conducted by the CO.
- Areas inspected by the CO.
- Each alleged violation—close up, then the surrounding area.

Note that the videotape is not a substitute for conventional photographs—it is a supplement. Each videotape must be identified as indicated above for conventional photographs.

Upon completion, all photographs and videotapes are to be immediately delivered to the (**COMPANY**) Safety and Health Manager, or to the Project Manager if more expedient.

8.7.6 Inspection Supplies and Equipment

Certain materials, equipment, and supplies are to be maintained on the jobsite or in a manner otherwise immediately available to the (**COMPANY**) Representative whenever needed in an inspection event. They must be maintained in condition to be immediately available to the site for the purposes of performing all documentation as required in this ***Safety and Loss Control Program Manual***. The inventory of these items should include at a minimum:

1. 35mm camera with flash capability. An instant camera with flash capability is acceptable only if a 35mm camera is not available. A camera with date and time capability is recommended. Such cameras are available to be signed out of the (**COMPANY**) home office, along with film and supplies.
2. Three to six packages of 24– or 36–exposure color print film.
3. Audiotape recorder. Dictation, pocket size is acceptable, but may prove to be inconvenient due to the small tape capacity. Regular cassette size is preferable.
4. Audiotapes totaling a capacity of (2) to (3) hours.
5. Flash items and batteries as required by the particular photographic and tape recording equipment.
6. 50′ tape measure.
7. 1′ ruler, or other device that can show relative sizes in photos.
8. All necessary personal protection equipment (hard hat, safety glasses, hearing protection, respirators, gloves, etc.).

Historically, audiotape recorders have not been permitted by OSHA, and may not be allowed.

8.7.7 Inspection Procedure

Upon first contact, the OSHA Compliance Officer should present his or her identification to the Project Manager or other senior (**COMPANY**) on-site official, and state the purpose of the visit. He or she should then request that an opening conference be held with company representatives from all contractors and subcontractors on the site, union stewards if any, and the safety officials of (**COMPANY**) and others on-site.

NOTE TO MANUAL PREPARER: Consider the following sample information very carefully, and confirm its applicability with respect to the intentions of your own actual company policy.

When the requested personnel have assembled, it must be determined if a warrant will be required from any of the individual contractors or subcontractors. (Ideally, this should have been determined at the start of the project via a survey of all contractors and subcontractors as to their intentions in this regard.) ***It is the policy of (COMPANY) that a warrant not be required***, but the various contractors and subcontractors and/or trade contractors on the site each have individual rights to make this determination for themselves.

If it is determined that a warrant will be required, the OSHA Compliance Officer must be notified then of the requirement before being allowed to proceed. Once the Compliance Officer has been so notified, no further action should be taken by the Compliance Officer until a properly executed warrant is presented. *Notify the (***COMPANY***) Safety and Health Manager and/or senior management immediately if any notice of required warrant has been given to the OSHA Compliance Officer.*

If a warrant is not required by any contractor or subcontractor, the OSHA Compliance Officer will begin the opening conference. In the opening conference, the OSHA Compliance Officer will:

1. Advise if the inspection is a general complaint inspection, target industry inspection, or other type of inspection.
2. State the approximate time that he or she expects to be on-site conducting the inspection, or performing any other activity.
3. Generally discuss the purpose of the OSHA act, its sanctions, and the authority invested in the OSHA Compliance Officer by the Act.
4. Approve members of the inspection party. Each employer has the right to representation, but the Compliance Officer has the right to object to the

representative for a valid cause. Disruptive conduct by the employer/employee representatives is cause for dismissal from the inspection party.

5. Request copies of company safety programs, any accident reports, and inspection surveys. When requested, these materials are to be made available to the Compliance Officer. If contract documents are requested, refer the Compliance Officer to the (**COMPANY**) Safety and Health Manager or to senior management.
6. Advise that a closing conference will be held at the conclusion of the inspection in order to advise of any alleged violations observed, and determine dates of correction.

During the inspection, all company representatives should proceed with the following general conduct:

1. Allow the OSHA Compliance Officer to lead the inspection.
2. Never harass or intimidate the OSHA Compliance Officer in any way.
3. (**COMPANY**) employees who are not necessary to the inspection are not to be allowed to attend, follow, or stay near the inspection party.
4. Access for inspection of areas that contain trade secrets may be denied to the OSHA Compliance Officer while such trade secret operations are ongoing. In such cases, advise the Compliance Officer of the condition. If the Compliance Officer then proceeds in the area, advise the Compliance Officer that you do not consider this to be part of the inspection. Keep detailed notes, photographs, and appropriate records of the event.
5. Record the specific route taken by the OSHA Compliance Officer. Note if the Compliance Officer talks to employees, if so, to whom, and note if he or she returns to any location that had been previously inspected.
6. If photographs are taken by the OSHA Compliance Officer, ask the Compliance Officer to describe the nature of the suspected violation. Record all remarks and relevant observations, and deliver to the (**COMPANY**) Safety and Health Manager upon the conclusion of the inspection.

8.7.8 The Closing Conference

At the completion of the inspection, the OSHA Compliance Officer will either conduct a general closing conference with all contractors and subcontractors in attendance, or will meet with each individual contractor and subcontractor.

A (**COMPANY**) representative should attempt to attend all meetings if any are being conducted individually. If allowed to attend, all information is to be recorded and included in the information delivered to the Safety and Health Manager.

If the OSHA Compliance Officer will not allow attendance at the individual meetings, immediately notify the (**COMPANY**) Safety and Health Manager and (**COMPANY**) senior management.

Do not admit responsibility for any alleged hazard or violation, nor offer to correct hazards created by others at any time. (Note that this conduct is not being advised with any intention to obstruct any investigation in any way. It is only intended to allow the company to preserve its legal rights.)

8.8 OSHA CITATIONS

As a result of an OSHA site inspection, it is possible that a citation and notice of monetary penalty may be issued by the OSHA inspector. Such issuances may be made while the OSHA inspector is still on site, or may be sent to the company subsequent to the inspection. If any OSHA citation or penalty notice is delivered to a company employee on any jobsite, the employee is to immediately deliver the issuance to the senior (**COMPANY**) jobsite supervisory or management staff on the site. That senior company individual must then:

1. Immediately forward copies of all materials received to the (**COMPANY**) Safety and Health Manager, including any completed copies of inspection records and photographs. If there is any question, forward all materials to the (**COMPANY**) home office.
2. If necessary, further instruction will be given to the jobsite field staff through the Safety and Health Manager or otherwise from the (**COMPANY**) Home Office.
3. Copies of any citations must be posted near the area cited for a safety or health violation. Failure to post citations as required is punishable by additional fines.
4. Such document notices must remain posted for (3) working days or until required corrections have been made. Note that each citation as it is issued should provide for the removal of assessed penalty figures with acceptable corrections. Be sure that this section of the citation is cut away *before* the citation is photocopied and posted on the site.

8.9 COMPANY REPORT OF OSHA INSPECTION

The sample Company Report of OSHA Inspection included in this section is to be used by company personnel to record the particulars of every visit by any OSHA representative to any (**COMPANY**) jobsite. It is to be completed by the senior on-site company employee as the (**COMPANY**) representative during any OSHA inspection.

Copies of the report must be immediately distributed to (**COMPANY**):

- Safety and Health Manager
- Project Manager
- Superintendent
- Senior Project Manager.

COMPANY LOGO	Construction Safety and Loss Control Program Manual	Page: 8-18	Appr:
	8. THE OCCUPATIONAL SAFETY AND HEALTH ACT (OSHA)	Revision Date:	

(Sample) Company Report of OSHA Inspection

Project: ______________________ No.: ________ Date: ____________ S M T W T F S
Location: ______________________________ Time: ____________ AM PM

Compliance Officer: Name: ______________________ Badge #: ____________
OSHA Office Address: ____

1. Did OSHA Compliance Office wait for the (**COMPANY**) Safety & Health Mgr.: ___ YES ___ NO
2. Was Opening Conference Held: ___ YES ___ NO
3. Name of (**COMPANY**) Representative: ______________________
4. Complete Description of Perceived Violation Area *(include all dimensions, slopes, proximity to other work and all other relevant information)*

____
____
____
____

5. Photographs taken by: ______________________ # Photos Taken: ______
6. Audio Tape Recordings taken by: ______________________
7. Videotape Recording taken by: ______________________
   Location/direction/description of videotape recording: ______________________
   ____
   ____
8. Were all (**COMPANY**) employees wearing hard hats and other appropriate personal protective equipment: ___ YES ___ NO
   If no, provide complete explanation: ______________________
   ____
   ____
9. Were barricades and other required area protection devices properly in place: ___ YES ___ NO
   If no, provide complete explanation: ______________________
   ____
   ____
10. Was a general Closing Conference held: : ___ YES ___ NO
11. Were individual Closing Conferences held: ___ YES ___ NO
12. If yes, was (**COMPANY**) representative allowed to attend: : ___ YES ___ NO

REMARKS: ____
____
____
____
____

(COMPANY) Representative:
Signed: ______________________
Print: ______________________

Distribution:
- ☐ Safety & Health Manager
- ☐ Project Manager
- ☐ Senior Project Manager
- ☐ CF

Sample Section 9

Environmental Health Hazards

Section Use

The form and content of sample Section 9 *Environmental Health Hazards* is designed as an example of company instruction to its employees. It is intended to provide for each employee:

- The concept of industrial hygiene, and how it is applied in the workplace.
- Basic knowledge about several of the more common and serious sources of hazardous materials, with explanation of their effects on the human body.
- Specific direction with respect to appropriate employee action when certain hazardous materials are discovered on a company jobsite.
- Guidance with respect to the preparation and maintenance of the various employee exposure records and medical records that may become necessary in the event of exposure of an employee to a toxic or hazardous substance.

Carefully consider these objectives during the review and editing process of the section material in order to customize the section content to the specific needs of your organization. Prepare final language that is consistent with decisions made by your own management in the determination of your own company program.

It is important to understand that safety rules and regulations change from time to time. Be sure to review the final draft of the section and the entire

program manual with appropriate safety, insurance, and legal professionals prior to issue and use. Periodically review your program with company management and all appropriate professionals to be certain that the company program remains current with all requirements.

COMPANY LOGO	Construction Safety and Loss Control Program Manual	Page: 9-1	Appr:
	9. ENVIRONMENTAL HEALTH HAZARDS	Revision Date:	

(Sample)

SECTION 9

Environmental Health Hazards

COMPANY LOGO	Construction Safety and Loss Control Program Manual	Page: 9-2	Appr:
	9. ENVIRONMENTAL HEALTH HAZARDS	Revision Date:	

9.1 INDUSTRIAL HYGIENE

The term "industrial hygiene" is one that applies to the identification and control of the various types of environmental health hazards that may be present or that may occur in the work place to which an employee may become exposed during the course of employment. It is the science of recognizing, evaluating, and controlling those environmental factors or stresses that may cause or contribute to health impairment, sickness, death, or even significant discomfort to employees or to others in the area's proximity. These factors may be of chemical, physical, biological, and ergonomic natures.

This section of the ***Safety and Loss Control Program Manual*** is intended to identify several concerns that are relatively common in the construction environment, and provide general guidance in procedures that may help correct problem situations as they may arise. It is by no means to be considered exhaustive. Rather, it is to be considered as an example of approaches to specific situations once they have been identified. If any area of question or concern arises on a jobsite or shop area, the (**COMPANY**) Safety and Health Manager, appropriate Project Manager, or (**COMPANY**) senior management must be contacted immediately upon observation of such condition.

9.2 RESPIRATORY PROGRAM

9.2.1 Introduction

This section is intended to present certain basic information regarding respiratory protection considerations of which each employee should be aware. Note, however, that the material is not to be considered as all inclusive in content or scope, but is presented as general information. Complete protection of the worker may include several elements of a personal protection program. Any element of such a program may at a given moment be more important than others for a given jobsite environmental condition.

Responsibility and authority for administration of any respiratory protection program is assigned to the Project Manager, who may delegate this authority to the Site Superintendent. If delegated, however, the Project Manager will remain responsible for assuring compliance with the complete program.

The responsible (**COMPANY**) employee is to contact the Safety and Health Manager before proceeding if there are any questions, or if specific assistance is required to confirm proper procedure.

9.2.2 Respiratory Hazards

Hazardous substances can enter the body in many ways. They include:

1. Inhalation into the lungs.
2. Absorption through the skin.
3. Ingestion into the gastrointestinal tract.

Of all possible manner of entry of hazardous or toxic substances into the body, the respiratory system is usually the fastest and most direct. It is the intimate association of the respiratory system with the circulatory system, combined with the need to constantly oxygenate tissue cells throughout the body, that are the characteristics that create this condition. The other methods are important to understand, to be sure, but this section focuses on the entry avenue that is the greatest risk to all employees.

In the evaluation of jobsite areas that may result in exposure of an employee to unsatisfactory respiratory conditions, it is important to understand that oxygen deficiency can occur in confined spaces by the displacement of air by other gases and vapors. It can also occur as direct result of the oxidation process. Such oxidation processes include not only fire, but rusting, action of aerobic bacteria, and degradation of organic material (such as giving off methane in an old dump area) as well.

The normal concentration of oxygen in fresh air is 20.6% by volume (the balance is mostly nitrogen—an inert gas as far as the body is concerned). Oxygen concentrations below 16% at sea level are considered harmful for human exposure, and will not even support combustion. At oxygen concentrations below 16%, physical collapse can occur without warning. Death can result within minutes.

Although 16% oxygen is generally considered the lower limit for safe human exposure, it is the partial pressure of oxygen that is the other significant factor in the level of overall risk. Under conditions of significantly reduced atmospheric pressure, for example, which occur at high altitudes, the low safe limit of oxygen is higher than 16%. In order to arrive at a standard that can be consistently applied throughout the country, OSHA currently requires a minimum value of not less than 19.5% by volume as of the time of this writing.

The atmosphere in a confined space may include particulate contaminants in various forms. In the classification of atmospheric particulate contaminants, the particle size is of particular significance. It is recognized, for example, that particles below ten (10) microns in diameter have a good chance of entering the respiratory system, and particles below five (5) microns are able to penetrate deep into the lung. The constant cleansing action of the regular breathing process usually removes particles of these sizes from the healthy lung. Excessive dust conditions, or diseased respiratory systems can, however, greatly reduce the effects of the normal cleansing process.

If it is ever suspected that any type of air contamination is present in any **(COMPANY)** jobsite location or if any jobsite area may possibly be considered to be a confined space as defined by OSHA and outlined in this section, *the Project Manager must be notified immediately*. (If the project manager is not available for any reason, the observing employee is to immediately notify the **(COMPANY)** Safety and Health Manager directly.) The project manger will then notify the **(COMPANY)** Safety and Health Manager. The project manager in consultation with the Safety and Health Manager as necessary will then determine if expert assistance will be necessary before allowing the work to proceed in the area.

NO WORK IS TO PROCEED IN ANY SUSPECTED AREA UNTIL PROPER PROCEDURE IS DETERMINED AND IMPLEMENTED, OR THE AREA IS CLEARED BY THE (COMPANY) PROJECT MANAGER OR SAFETY AND HEALTH MANAGER.

There may be situations occurring on jobsites that may be beyond the ability of the Safety and Health Manager to resolve. In such cases, experts may be retained to deal with the specific job condition. Determinations will be made with respect to the following:

- Hazard assessment methodology
- Hazard control procedures
- Selection of environmental control equipment and personal protection devices
- Proper training in the use of all equipment.

Workers should never be assigned to any operations requiring respiratory protection until a physician has determined that they are physically and psychologically able to perform the planned work while using the respiratory protective equipment as it is intended. These worker conditions will also be reviewed by available qualified experts who are responsible for considering the specific jobsite condition.

9.3 CONFINED AREA ENTRY

9.3.1 General

A "confined" area is not necessarily one that restricts movement or operation, but can be. A confined area or space as considered for the purposes of this section is any area in which the air contamination is not prevented or removed by natural ventilation. In the event of any exposure by an employee to such an environment, there is always a possibility of the presence of combustible or toxic gases or particulates and/or a reduced level of oxygen.

Such health hazards that are not visible, have no odor, are not heard, or cannot be felt too often represent some of the most immediate deadly risks to employees. The clandestine nature of these hazards should cause each employee who works in such confined areas to take these risks very seriously.

Of the atmospheric conditions that may constitute health hazards to employees on jobsites, the three most common as identified by OSHA are the following:

1. *Toxic vapors and gases.* These are atmospheres that contain substances that can cause serious health impairment, injury, or death. Asphyxiates and irritants are included in this category. It is important to note that the particular concentrations of such contaminants are not necessarily the significant factor, because certain substances can cause their adverse affects even at low concentrations.
2. *Combustible vapors and gases.* These are atmospheres that contain combustible substances. These substances may explode or ignite if a source of ignition is introduced into the environment. Often it is the fact that these substances are in a gaseous, rather than liquid state, that causes or contributes to their explosive nature.
3. *Oxygen deficiency.* These are atmospheres that contain less than 19.5% oxygen. Refer to related sections of this *Safety and Loss Control Program Manual* for additional discussion. Normal fresh air contains 20.9% oxygen. Combustion is not possible below concentrations of oxygen below 16%. Actual detriment to humans may occur at oxygen concentrations less than 19.5% at sea level, but 19.5% is the current OSHA standard.

9.3.2 Susceptible Areas

Certain types of areas are more susceptible to allowing these kinds of conditions to occur than others, and some are much more obvious than others. Every employee should proceed with the expectation that any one or more of the atmospheric conditions described above may occur in areas such as:

- Aqueducts,
- Ductwork, building,
- Ductwork, underground,
- Elevator machine rooms,

- Excavations,
- Manholes,
- Pool pump rooms and pump houses,
- Piping (large diameter),
- Rooftop mechanical penthouses,
- Sewers,
- Subcellars,
- Storage tanks,
- Utility vaults, above ground,
- Utility vaults, below ground,
- Water filtration plants,
- Water storage areas,
- Similar types of construction.

These are some of the more obvious types of areas that are at higher risk of allowing dangerous atmospheric conditions. Before entry is made into any area suspected to be a confined area that may contain hazardous atmospheres, always proceed with the assumption that a hazard exists, and that purging or ventilation will be required.

9.4 CARBON MONOXIDE

9.4.1 General

Carbon monoxide is a gas that is flammable, toxic, non-irritating, tasteless, odorless, colorless, and heavier than air. It is formed by the incomplete combustion of materials such as oil, gasoline, natural gas, and manufactured gas.

Examples of conditions on a construction site that can cause the presence and build-up of carbon monoxide are temporary heaters or gasoline motors used in confined areas or depressed areas where men and women are working, or any internal combustion engine allowed to operate in a closed or poorly ventilated area.

If carbon monoxide is inhaled, the respiratory process allows it to combine with the hemoglobin in the blood, thereby excluding oxygen from the body tissues. This process is the cause of carbon monoxide poisoning, and ultimately results in death.

Some of the more obvious symptoms of carbon monoxide poisoning are headache, dizziness, shortness of breath, nausea, and muscular weakness.

NOTE TO MANUAL PREPARER: Verify the most current requirements of OSHA and any other controlling legal standard, and confirm the availability of carbon monoxide testers before finalizing the remarks of the following sample section.

9.4.2 Work Area Procedures

It is the policy of (**COMPANY**) that in every case on each construction project where any piece of equipment being used or intended to be used that is to discharge carbon monoxide into the area where any workers or other employees are required to work or to otherwise be present for any reason, (**COMPANY**) senior project management shall:

1. Ensure that all areas are properly ventilated. Provide positive ventilation as necessary to maintain carbon monoxide levels below acceptable and legal limits at all times.
2. Arrange to have the work area tested in order to determine the levels of carbon monoxide in the atmosphere at a minimum of (3) times each eight (8) hour period. Carbon monoxide testers are available from the (**COMPANY**) Safety and Health Manager for direction prior to proceeding.
3. Test several locations within each suspected area. Be sure to include tests taken at the breathing height(s) of the worker(s).
4. Record the results of each test(s). Include the testing date and time(s), note the individual(s) who are taking each test, and summarize the results of each test.
5. Keep a copy of each test record in the jobsite files, and distribute one copy of each test to the (**COMPANY**) Safety and Health Manager.

9.5 LEAD

9.5.1 General

Lead can be present in various forms on a jobsite. Most often, it will be in more obvious solid form, but it can also be present as an atmosphere particulate, or even a vapor or gas. Common sources of lead exposure can occur during the cutting of lead-based metals, and in the cleanup and removal of lead-based paints, and in welding work.

If the presence of lead is suspected in any area, and the area is to be tested for lead, it is important to verify current permissible exposure limits at that time. At the time of this writing, the permissible exposure limit of particulates is 0.05 milligrams per cubic meter; the vapor exposure varies, because inhaled vapors are accumulative within the body organs. Confer with the **(COMPANY)** Safety and Health Manager to arrange for lead testing.

Symptoms of exposure to lead may include abdominal pain, colic, anemia. Prolonged and intense exposure can result in gingival lead line tremors, and paralysis. There may be other effects, as well.

9.5.2 Work Area Procedures

Respect lead for the damage that exposure to it may cause. When working in and around areas known or suspected to contain lead, observe the following:

1. Whenever possible, work in lead areas shall be adequately ventilated. This is defined as a moderate airflow across the face.
2. When such ventilation is not possible, approved filter masks or positive pressure hoods shall be used.
3. Food or drink are not to be taken into areas where the possibility of lead exposure exists.
4. Disposable coveralls, head coverings, and gloves shall be worn in contaminated areas, and disposed of after each use.

Contact the **(COMPANY)** Safety and Health Manager before proceeding in every instance, and if there is any question.

9.6 ASBESTOS

NOTE TO MANUAL PREPARER: The self explanatory remarks of the sample section that follows anticipate a company that is not currently and not interested in becoming involved directly in any asbestos removal or abatement on construction jobsites. As such, consider the language in this context, and modify as appropriate to accommodate the specific decisions made by your own company in this regard.

9.6.1 General

Asbestos is a building material that occurs in several forms in construction sites. It may be present in several types of areas of existing buildings, and may be uncovered during renovation operations in buildings and in underground utilities in structures built prior to 1973.

Examples of possible locations where asbestos may be present include:

- Spray-on fireproofing at steel beams, columns, and decking,
- Pipe or boiler insulation,
- Ceiling tiles,
- Wall and ceiling plaster,
- Floor tiles,
- Pipe insulation in underground ducts,
- Other various materials used to fireproof and insulate.

Asbestos is a carcinogen. Exposure to asbestos may not lead to immediate detectable symptoms, but can ultimately result in various forms of cancer or other disease that may manifest symptoms up to many years after the exposure incident.

This especially insidious nature of asbestos exposure should command adequate respect from all workers who are potentially exposed to its effects.

9.6.2 Asbestos Presence

In situations where the presence of asbestos is known and has been tested, (**COMPANY**) is entitled to receive a properly certified test analysis by a qualified testing agency, certifying that the concentrations of asbestos in the environment are within acceptable levels and legal limits. Even if (**COMPANY**) is not intending to work in close proximity to the asbestos area, we are still entitled to, and should receive a copy of any such report.

Without this type of complete and proper documentation from a qualified and approved source, ***no (COMPANY) employee is to enter any area known to contain such asbestos.*** There may be occasions where company employees may be allowed into such areas while using proper equipment and having received proper training. This situation, however, will be a very special case. It will only be done if specifically cleared through the (**COMPANY**) Senior Project Manager *and* the Safety and Health Manager.

NO COMPANY EMPLOYEE SHALL EVER INTENTIONALLY TOUCH, DISLODGE, REMOVE, HANDLE, OR DISPOSE OF ANY MATERIALS KNOWN OR SUSPECTED TO CONTAIN ASBESTOS. A POSSIBLE EXCEPTION MAY BE VINYL ASBESTOS FLOOR TILE (VAT), BUT THEN ONLY IN SPECIAL CIRCUMSTANCES THAT HAVE BEEN GIVEN DIRECT CONSIDERATION BY THE COMPANY SAFETY AND HEALTH MANAGER. ANY WILLFUL VIOLATION OF THIS POLICY WILL BE GROUNDS FOR DISCIPLINARY ACTION.

If asbestos presence is suspected in any area not previously identified as containing asbestos, the suspecting employee is to advise the (**COMPANY**) project manager immediately. The project manager should then:

1. Inspect the area to confirm the apparent condition as described by the observing employee.
2. If the area had previously been inspected and tested, and found to be free of asbestos, advise the employee—*noting the copy of the associated test report that confirms the lack of asbestos.*
3. If the project manager is not in possession of such a report, but believes that one exists, *immediately secure a copy of the report from the responsible party* (owner, general constructor, construction manager, etc.).
4. If the area is indeed suspicious, and has not been previously tested, proceed as described in section 9.6.3 and 9.6.4 below.

9.6.3 Asbestos Procedures and Company Policy

Due to the risk of serious long term effects of exposure inherent in the removal and disposal of asbestos, the following (**COMPANY**) policy must be clearly understood by all employees:

1. No company employee shall ever intentionally touch, dislodge, remove, handle, or dispose of any materials known or suspected to contain asbestos. A possible exception may be Vinyl Asbestos Floor Tile (VAT), but then only in special circumstances that have been given direct consideration by the company Safety and Health Manager, and after specific procedures have been established and approved by (**COMPANY**) project management.
2. (**COMPANY**) will not contract or subcontract for the removal of asbestos on any project unless the asbestos removal will be performed by a qualified and certified asbestos removal specialty contractor who is licensed, bonded, and with insurance in coverages, amounts, and named insured that are satisfactory to (**COMPANY**).
3. If no such qualified, licensed specialty contractor is available for any reason, or if the specialty contractor is unable to provide either the necessary bonds or insurance coverage, (**COMPANY**) will decline to commit itself to the asbestos removal portion of the contract.

4. If asbestos materials are ever discovered or suspected during the course of performing any work, *all work in the immediate vicinity shall immediately be suspended*, and *all (**COMPANY**) employees shall immediately leave the area.* Project supervisory staff, the client, and (COMPANY) project management and supervisory staff shall immediately be notified. Work in the affected areas shall not resume until the material is either determined not to be asbestos, or the asbestos has been properly encapsulated and/or removed and disposed of by others in a legal, properly insured manner.

Under no circumstances will any (**COMPANY**) employee ever knowingly or intentionally enter or remain in areas containing asbestos, remove or attempt to remove any amount of any form of asbestos, dispose of any materials containing asbestos, or allow any other (**COMPANY**) employee to do the same.

9.6.4 Required Action Upon Discovery of Asbestos

If at any time any (**COMPANY**) employee encounters any material on a jobsite that is suspected of being asbestos, the employee is to immediately stop work in the area and report the condition at once to his or her immediate supervisor. The senior person on the jobsite will then:

- Take immediate steps to prevent any other persons from entering the area.
- Depending on the job condition, determine if the area should be enclosed with plastic before ceasing all operations in the affected area. If so, make immediate arrangements to enclose the area.
- Immediately contact (**COMPANY**) senior management and the (**COMPANY**) Safety and Health Manager for further direction.
- (**COMPANY**) senior management will contact the project owner and architect, notify of the condition, and immediately confirm all circumstances in writing.

No further work in the area will proceed until the proper procedure has been established and agreed to by (**COMPANY**) senior management.

9.7 EMPLOYEE EXPOSURE RECORDS AND MEDICAL RECORDS

NOTE TO MANUAL PREPARER: The sample information that follows has been prepared as an example of how a company might address the respective legal

requirements discussed. It has been prepared in consideration of associated laws and OSHA regulations in effect as of the time of original material preparation. It is important to recognize that such laws and regulations may change from time to time, and are subject to interpretation during the preparation of such material. The preparer should secure the most current information available for use in the preparation of any material that is eventually used in any actual company Safety and Loss Control Program Manual.

9.7.1 Section Purpose

The purpose of this section is to generally describe for all **(COMPANY)** employees their rights of access to relevant employee exposure records and medical records. It also attempts to summarize for each employee certain rights of the office of the Assistant Secretary of the Department of Labor of access to these records as are necessary to allow the office to fulfill its responsibilities under the Occupational Safety and Health Act.

Access by employees, their representatives, and the Assistant Secretary is considered necessary in order to allow and encourage ongoing improvements in the detection and treatment of occupational illness, and to identify and correct the causal conditions of injuries and illnesses, and to improve prevention of occupational disease. The company is responsible for compliance with the procedures described in this section, but the actual activities involved in compliance with the rules of access to medical records may be conducted on behalf of the employer by a physician or other health care professional who is directly responsible for the subject employee medical records.

It is important to note that nothing in this section or in the complete ***Safety and Loss Control Program Manual*** is intended in any way to affect existing moral and legal principles and obligations concerning the confidentiality of employee medical records, or any other aspect of the employer–employee relationship. Similarly, nothing is intended to alter or otherwise affect any legal rights or obligations regarding the preservation of trade secrets and similar information.

This section applies to the company to the extent that the company makes, maintains, contracts for, or has access to employee exposure or medical records, or analysis of such records as they pertain to employees who have been exposed to toxic substances or potentially harmful physical agents, and to all such employee exposure and medical records.

It is the company's responsibility to ensure that the employee medical record preservation and access requirements described in this section are complied with, regardless of the actual manner in which such records are actually prepared, stored, distributed, and maintained.

9.7.2 Definitions

The definitions of this section apply to the use of these terms in the various laws and regulations relating to employees' access to their own exposure records and medical records. These definitions are taken from the latest OSHA materials available as of the time of material preparation. These are informal guidelines only, and are not intended to be considered as necessarily definitive in the legal sense. Check with the company Safety and Health Manager in the event of any question.

Access: Right and opportunity to examine records and copy, by whatever means available.

Analysis: With reference to analysis using employee exposure or medical records, any compilation of data, or any research, statistical, or other study based at least in part on information collected from individual employee exposure or medical records, or information collected from health insurance claim records, provided that either the analysis has been reported to the employer or no further work is being done on said analysis.

Authorization (Also: Consent, Written Consent, and Specific Written Consent): A written notice of authorization that contains the information listed below. Note that such an authorization may be withdrawn or revoked by the employee at anytime, and usually will not remain in effect beyond one year after the date of the authorization:

a. Date of the authorization.
b. Description of the medical information that is being authorized to be released.
c. Name of the individual or organization that is authorized to release the medical information.
d. Name of the designated representative that is authorized to receive the medical information.
e. General description of the purpose for the release of the medical information.
f. Date or condition upon which the written authorization will expire (if less than one year).
g. Statement that the release of future medical information that is not in existence on the date of the authorization is expressly authorized *only if the employee specifically intends this situation*. Otherwise, silence to this condition will result in *no* authorization to release such future information.
h. Name and signature of the employee authorizing the release of medical information.

COMPANY LOGO	Construction Safety and Loss Control Program Manual	Page: 9-14	Appr:
	9. ENVIRONMENTAL HEALTH HAZARDS	Revision Date:	

Employee: A current employee, former employee, or new employee. In the case of a deceased or legally incapacitated employee, the employee's legal representative.

Employer: Former employer, current employer, or a successor employer.

Exposure: Subjection of an employee to a toxic substance or harmful physical agent through any route of entry into the body, such as inhalation, ingestion, skin contact or absorption. It may include past and potential exposure, but does not include situations where the employer can demonstrate that a questionable suspected substance that is the subject of the exposure allegation is not or was not actually used, handled, stored, generated, or otherwise present in the respective area(s) in any manner that is or was different from similar situations in typical non-occupational settings.

Exposure Record (Employee): A record containing information concerning employee exposure to toxic substances, harmful physical agents, or otherwise hazardous materials, including: (a) environmental monitoring or measuring; (b) biological monitoring results which consider the absorption of a substance or agent by body systems; (c) Material Safety Data Sheets (MSDSs); and (d) in the absence of the above, any other information in any form of medium that reveals or may reveal the identity or circumstances of a toxic substance or harmful physical agent.

Harmful Physical Agent: See "Toxic Substance."

Record: Any item, collection, or grouping of information regardless of the form or process by which it is maintained (paper, computer diskette, microfilm, etc.)

(Designated) Representative: Any individual or organization to whom an employee gives written authorization to exercise a right of access. In the case of a deceased or legally incapacitated employee, that employee's legal representative.

Medical Record (Employee): A record regarding the specific health status of an employee which is generated and/or maintained by a physician, nurse, or other health care professional.

Toxic Substance: Any biological agent, chemical substance, or physical stress (noise, heat, cold, vibration, repetitive motion, radiation, hyperbaric pressure, etc.) which is:

a. Regulated by federal law or rule to a related hazard to health;
b. Listed in the latest printed edition of the *National Institute for Occupational Safety and Health* (NIOSH) *Registry of Toxic Effects of Chemical Structures* (RTECS);

c. Has a Material Safety Data Sheet available to the employer indicating that the material may pose a hazard to human health;

d. Has yielded positive evidence of an acute or chronic health hazard in human, animal or other biological testing conducted by or known to the employer.

9.7.3 Preservation and Maintenance of Records

It is the company's responsibility to ensure the preservation and retention of records for the duration of employment, plus the minimum times and conditions listed below, unless a different time is provided by a specific OSHA standard:

1. Analysis using exposure or medical records: (30) years
2. Employee exposure records: (30) years
   Exceptions:
   (a) Background data to environmental (workplace) monitoring or measuring need only be maintained for (1) year, as long as the result and methodology are retained for (30) years, and
   (b) Material Safety Data Sheets and certain records that identify the substance or agent do not have to be retained, so long as some adequate record of the identity and use of substance is retained for (30) years.
3. Employee medical records: (30) years
   Exception: Health insurance claim records maintained separately from the employer's medical program may not be required to be retained for any specific period.

The records of this section will be preserved directly or arranged to be preserved in their proper states by the (**COMPANY**) Human Resources Department. Contact the company Human Resources Manager for additional information.

9.7.4 Employees' Access to Records

NOTE TO MANUAL PREPARER: As of the time of this writing, government regulations do apparently allow the employer to charge an employee reasonable costs of copying records under certain repetitive conditions. These conditions were not addressed in this sample section as an example of a company policy. Your final decision in this regard may ultimately be different as you may determine for yourself.

In every case where any employee or the employee's designated representative requests access to a record and/or requests a copy of such records, the company shall:

1. Ensure that access is provided in a reasonable time, place, and manner, but in no event later than fifteen (15) calendar days after the request for access is made.
2. Within the period of time described in (1), assure that either:
   a. A copy of the record is provided without cost to the employee or representative if a copy is requested,
   b. The necessary copying facilities are made available without cost to the employee or representative for copying the record, or
   c. The record is loaned to the employee or representative for a reasonable period of time to enable a copy to be made.

It is the responsibility of the **(COMPANY)** Human Resources Manager to satisfy the requirements of this section of the *Safety and Loss Control Program Manual* on a continuing basis.

9.7.5 Employer's Notices to Employees

Upon an employee's initial employment, and at a minimum of each year thereafter, the company is responsive to inform all employees who may be at risk of exposure to toxic substances, hazardous materials, or harmful physical agents: (a) the existence, location, and availability of all records regarding employee safety, health, or exposure; (b) the individual and/or department responsible for maintaining such records and allowing access to them, and; (c) every employee's rights of access to all such records.

The company will make available to its employees a copy of all available relevant OSHA and other standards as may be applicable. The company will distribute any information materials concerning such standards which are made available to the company by the Assistant Secretary or Labor for Occupational Safety and Health, as they are made available or as otherwise required.

Further information regarding the items discussed in this section is available through the **(COMPANY)** Human Resources Manager and the Safety and Health Manager.

9.7.6 Transfer and Disposal of Records

NOTE TO MANUAL PREPARER: The first item below pertains to companies that have been in business and otherwise have been in a position to have the type of employee exposure and medical records discussed in this sample section for at least thirty years. If this is not the case, you may consider omitting that time from your own actual Safety and Loss Control Program Manual until such time as those conditions change.

COMPANY LOGO	Construction Safety and Loss Control Program Manual	Page: 9-17	Appr:
	9. ENVIRONMENTAL HEALTH HAZARDS	Revision Date:	

1. Where an employer regularly disposes of records that are required to be preserved for (30) years, the employer may, with at least three (3) months notice, notify the Director of The National Institute for Occupational Safety and Health (NIOSH) on an annual basis of the records intended to be disposed of in the coming year.
2. If the company should ever cease to do business, it will be responsible to transfer all records that are the subject of this section to any successor employer. The successor employer will then be responsible to receive and maintain these records.
3. If the company should ever cease to do business and there is no successor employer, the company will be responsible to notify all affected employees of their rights of access to records at least three (3) months prior to the cessation of the employer's business.
4. If the company should ever cease to do business and there is no successor employer, or if the company for any reason ever intends to dispose of any records required to be preserved for a continued period of time, the company will be responsible to:
   a. Transfer the records to the Director of The National Institute for Occupational Safety and Health (NIOSH) as may specifically be required by an occupational safety and health standard, or
   b. Notify the Director of NIOSH in writing of the intended disposal of records at least three (3) months prior to the disposal of such records.

It is the responsibility of the Human Resources Manager to comply with the requirements of this section, and to remain current with any changes in laws and regulations affecting the transfer and disposal of employee records.

9.7.7 Employee Right to Know Laws and Considerations

Employee Right to Know Laws, information regarding such laws, and responsibilities of employers with respect to such laws and considerations change often. In some cases, there may be confusion among regulations which can sometimes result in conflicts. There has been, for example, debate over whether the Federal OSHA standard will preempt all other regulations, or whether more restrictive regulations contained in a state and local regulation are in effect and must be complied with in the case of apparent conflict with federal standards.

If specific detailed information is needed on a particular situation, Arrange to secure the most up to date information and interpretations on its use directly from the State agency involved. Such research should be conducted through the (COMPANY) Human Resources Department. The (COMPANY) Safety and Health Manager is available to assist every company employee in this regard.

Sample Section 10

(COMPANY) HAZARD COMMUNICATION PROGRAM

Section Use

The form and content of Sample Section 10 *(***COMPANY***) Hazard Communication Program* is designed as an example of company instruction to its employees with respect to the material presented. It is intended to provide for each employee:

- A complete description of a Hazard Communication Program
- Instruction in the preparation of a *separate* Hazard Communication Manual, and how this safety and loss control program should correlate with it.
- Uses of the various hazard communication materials.

The sample section that follows is an example of a company safety and loss control program approach to the objectives indicated above. The sample language and forms are to be considered in the context of your own company and its manner of operation.

The preparer of the company's actual Safety and Loss Control Program Manual must become thoroughly familiar with the hazard communication process, the composition of the Material Safety Data Sheets (MSDSs), and the construction and use of a separate Hazard Communication Manual. This sample section is general information that will guide the preparer to the proper areas of concern.

Carefully consider these objectives during the review and editing process of the section material in order to customize the section content to the specific needs of your organization. Prepare final language that is consistent with decisions made by your own management in the determination of your own company program.

It is important to understand that safety rules and regulations change from time to time. Be sure to review the final draft of the section and the entire program manual with appropriate safety, insurance, and legal professionals prior to issue and use. Periodically review your program with company management and all appropriate professionals to be certain that the company program remains current with all requirements.

COMPANY LOGO	Construction Safety and Loss Control Program Manual	Page: 10-1	Appr:
	10. HAZARD COMMUNICATION PROGRAM	Revision Date:	

(Sample)

SECTION 10

(COMPANY) Hazard Communication Program

COMPANY LOGO	Construction Safety and Loss Control Program Manual	Page: 10-2	Appr:
	10. HAZARD COMMUNICATION PROGRAM	Revision Date:	

10.1 GENERAL COMPANY POLICIES

This section is intended to comply with the OSHA Hazard Communication Standard, Title 29 Code of Federal Regulations 1926.59. Specifically, the company intends to achieve this compliance by providing for each (**COMPANY**) employee the following:

- A complete list of hazardous chemicals used by the company and likely to be used by the company on any company jobsite,
- Description and use of Material Safety Data Sheets (MSDSs),
- Assurance that all containers are properly labeled, and
- Appropriate training in the handling and safe use of hazardous chemicals that are necessary in the performance of company work.

The company hazard communication (HAZ-COM) program applies to all work operations within the company where any employee may be exposed to a hazardous substance either under normal working conditions or during emergency situations.

The Safety and Health Manager is the HAZ-COM Program Coordinator. He or she is the (**COMPANY**) representative who has overall responsibility for the program's implementation. The Safety and Health Manager will review and update the program as necessary. Copies of the complete (COMPANY) ***Hazard Communication Program Manual***, as discussed in Section 10.2, may be obtained from the Safety and Health Manager at any time.

Under the HAZ-COM program, (**COMPANY**) employees will be informed of:

- The form and content of the Hazard Communication Standard,
- The hazardous properties of chemicals with which employees may work or come into contact,
- Procedures or safe handling of the various chemicals and materials required in the performance of their responsibilities,
- Measures of personal protection that can be taken against these chemicals, and
- The hazards associated with the performance of certain tasks that are not commonly experienced by employees.

10.2 (COMPANY) HAZARD COMMUNICATION PROGRAM MANUAL

NOTE TO MANUAL PREPARER: Consider the text of this sample Section as instruction regarding the preparation and deployment of the separate Hazard Communication (HAZ-COM) manual. The complete HAZ-COM manual should be distributed to everyone who requires its information and to everyone wishing to review it for any reason. Keep the HAZ-COM manual up to date at all times with Material Safety Data Sheets (MSDSs) for new materials as they are used by the company.

The complete **(COMPANY)** ***Hazard Communication Program Manual*** is a separate manual prepared and maintained by **(COMPANY)** that contains a reiteration of this section of the ***Safety and Loss Control Program Manual*** as its introduction, along with the actual copies of all Material Safety Data Sheets (MSDSs) for each of the products as listed in the following sections, and as supplemented to keep the list current. The MSDSs are for all products used by the company either on a routine basis, occasionally, or on a special case as required by a particular project.

Each copy of the HAZ-COM manual is to be maintained and kept current by the Safety and Health Manager. A copy of the current HAZ-COM manual is to be kept on each **(COMPANY)** jobsite. They are to be available to all **(COMPANY)** employees at any time from every Project Manager and from the **(COMPANY)** Safety and Health Manager.

The HAZ-COM manual is an integral part of the **(COMPANY)** safety and loss control effort, and must be referred to and used by every **(COMPANY)** employee in every relevant capacity on every **(COMPANY)** jobsite.

On multi-employer work sites where there exists any possibility of exposure of other employees to chemicals or substances used by **(COMPANY)**, sharing of the relevant Hazard Communication Program information is required. On certain projects, project management may be required to submit the specific MSDSs to the client to be retained on the jobsite as part of the client's safety program. On such projects, that effort will be considered by **(COMPANY)** as redundant to the HAZ-COM procedures required by this section of the company *Safety and Loss Control Program Manual*. In no case will any such separate procedures be allowed to replace the procedures required herein.

Contact the **(COMPANY)** Safety and Health Manager for additional information on use and availability of the HAZ-COM manual.

10.3 LIST OF HAZARDOUS CHEMICALS

The Safety and Health Manager will make a list of hazardous chemicals and related work practices used within the company, and will update the list as necessary. The list of chemicals identifies all the chemicals used at the (**COMPANY**) yard, within the shop area, and on all (**COMPANY**) jobsites.

The list is posted in the employee lounge, and a master list will be maintained by and is available from the Safety and Health Manager. The current list is also included in the company HAZ-COM manual described in Section 10.2 above, and in section 10.9 below.

NOTE TO MANUAL PREPARER: Sample Section 10.9 below is intended to include a list of all chemicals and substances that are used by the company as described, and for which an MSDS has been secured. Be sure that your own list is complete.

10.4 MATERIAL SAFETY DATA SHEETS (MSDSs)

Material Safety Data Sheets (MSDSs) will provide each employee with the specific information on chemicals that (**COMPANY**) employees may use or come in contact with during the course of their employment. The Safety and Health Manager will maintain the (**COMPANY**) ***Hazard Communication Program Manual***, which is a three-ring binder containing an MSDS on every substance on the list of hazardous chemicals included in Section 10.9 of this *Safety and Loss Control Program Manual*.

Each respective MSDS will be a fully completed OSHA Form 174 or its equivalent. The Safety and Health Manager will ensure that the list of MSDSs in the employee lounge is maintained and current.

The Safety and Health Manager is responsible for acquiring and updating all MSDSs. He or she will contact the chemical manufacturer vendor if additional research is necessary, or if an MSDS has not been provided within the initial shipment of the respective materials. All new procurements for the company must be cleared by the Safety and Health Manager. A master list of MSDSs is available at any time from the Safety and Health Manager.

OSHA specifies the information to be included on an MSDS, but does not prescribe the specific format for an MSDS.

OSHA requires that each MSDS must include the following information at a minimum:

1. Chemical identity.
2. Hazardous ingredients.
3. Physical and chemical characteristics.
4. Fire and explosion hazard data.
5. Reactivity data.
6. Health hazards.
7. Precautions for safe handling and use.
8. Control measures.

10.5 LABELS AND OTHER FORMS OF WARNING

The Safety and Health Manager will ensure that hazardous chemicals are maintained in their original labels as required by OSHA, and updated as necessary.

Each label must include:

- The chemical identity
- Appropriate hazard warning
- Name and address of the manufacturer or other responsible party

The Safety and Health Manager will refer to the corresponding MSDS in order to assist **(COMPANY)** employees in verifying label information.

At all times, chemicals will be maintained and transported in their original packaging that includes the labeling described above. If chemicals are transferred from a labeled container to a portable container *intended for immediate use*, no labels are required on the portable container.

10.6 PERFORMANCE OF NON-ROUTINE TASKS

When **(COMPANY)** employees are required to perform any tasks that may be hazardous that are not routinely performed by the company or the employee, a special training session will be conducted to inform affected and potentially affected employees regarding the hazardous chemicals to which the employee might be exposed, and the proper precautions necessary to reduce or eliminate the possibility of exposure.

Any employee being asked or directed to perform any task that he or she is unfamiliar, and that may involve possible exposure to a chemical, should immediately inform the employee's supervisor. The supervisor must then arrange for relevant training prior to requiring the employee to perform the task. It is the Safety and Health Manager's responsibility to either conduct the required training, or to arrange for qualified personnel to conduct the training.

10.7 (COMPANY) EMPLOYEE TRAINING

10.7.1 General

Everyone who works with or is potentially exposed to hazardous chemicals will receive initial training by the Safety and Health Manager on the hazard communication Standard and the safe use of those hazardous chemicals. Whenever a new hazard is introduced, additional training will be provided.

The regular safety meetings as more completely described in Section 4 of this ***Safety and Loss Control Program Manual*** will be used to review the information presented in the initial training. Foremen and other supervisors will be trained regarding the hazards and the purposes of the various protective measures, in order that they will be available to answer any questions from employees and be able to provide daily monitoring of safe practices.

If there are any questions regarding hazardous chemicals, every employee is to contact his or her immediate supervisor before any contact is made with the chemicals.

Training will emphasize:

- The summary of the Hazard Communication Standard and this written program.
- The chemical and physical properties of each hazardous material (such as flash point, reactivity, etc.), and methods that can be used to protect against the release of or other exposure to chemicals (including chemicals in unlabeled pipes).
- The physical hazards of chemicals (such as potential for fires, explosion, etc.).
- The health standards, including warning signs and symptoms associated with exposure to chemicals, and any medical conditions known to be aggravated by exposure to the chemical.
- All procedures to protect against exposure to hazards (such as personal protective equipment required, proper use and maintenance, work practices or methods to ensure proper use and handling of chemicals, and procedures for emergency response).
- The safe work procedures to be followed by each employee in order to assure protection when cleaning hazardous chemical spills and leaks.

- The locations of the company HAZ-COM manuals, including all MSDSs, instruction on reading and interpreting the information on both labels and MSDSs, and how employees may obtain additional hazard information.

The Safety and Health Manager or his or her designee will review the company's employee training program and advise on training or retraining needs. Retraining is required only when a hazard changes or when a new hazard is introduced into the work place, but it is the policy of (**COMPANY**) to provide training on an ongoing basis in the safety and loss control program's Safety Meetings in order to enhance the effectiveness of the total Program.

As part of the assessment of the training program, the Safety and Health Manager will obtain input from the employees regarding the training that they have received and their suggestions for improving any part of the entire training effort.

The most important aspects of training with respect to chemicals and the HAZ-COM standard are to ensure that (**COMPANY**) employees:

- Are aware when they are exposed to hazardous chemicals, and of the conditions of such exposure.
- Have the ability to read and use package labels and Material Safety Data Sheets.
- Follow all appropriate protective measures and company procedures as established.

As part of the ongoing OSHA inspection and enforcement efforts, it can be expected that OSHA Compliance Officers will talk with company employees specifically in order to determine if:

- They have received training with respect to chemicals identification and the Hazard Communication standard;
- They are aware that they are exposed to hazardous chemicals;
- They are aware of the locations in which they can obtain specific information on labels and MSDSs.

10.7.2 Training Program Outline

This section provides a guide to topics and format to be used by the Safety and Health Manager or his or her designee in the ongoing MSDS training of (**COMPANY**) employees.

COMPANY LOGO	Construction Safety and Loss Control Program Manual	Page: 10-8	Appr:
	10. HAZARD COMMUNICATION PROGRAM	Revision Date:	

1. **Introduction**

   a. Reasons for training
      i. Hazard Communication Standard
      ii. Providing for the employees' health and safety
   b. Description of the HAZ-COM standard requirements
      i. Material Safety Data Sheets
      ii. Available of MSDSs to all employees
      iii. Labeling requirements of every container
      iv. Provide a written program for the use of all company employees
      v. Provide information updates to all employees on a regular basis
      vi. Provide relevant training to all employees on the safe use and handling of chemicals

2. **Information to employees**

   a. Location(s) of posted information
      i. List of HAZ-COM standard requirements
      ii. List of operations that involve hazardous chemicals
      iii. List of locations where data sheets and copies of the HAZ-COM manual can be found
      iv. List of all hazardous chemicals in all office, shop, and jobsite locations
      v. Information regarding monitoring devices and/or alarms that are in use, as well as the methods and observations that may be used to detect the presence or release of hazardous chemicals

3. **Material Safety Data Sheets**

   a. Locations
   b. Reading and understanding Material Safety Data Sheets
      i. Reviewing properly completed data sheets as examples

4. **Review of all chemicals that are in each company work area. Highlighting of their hazards, detection methods, and employee protection methods**

   a. Chemical name
      i. Hazards
         1) Flammability
         2) Reactivity
         3) Toxicity by contact
         4) Toxicity by inhalation
         5) Cancer suspect?

6) Reproductive problems?
7) Irritating?
8) Causes burns?
9) Other hazards?

ii. Detecting the presence of the chemical
1) In the air?
2) On the skin?
3) By alarms and monitoring devices?
4) What level of chemical is hazardous?
5) What symptoms indicate overexposure?

iii. Preventing overexposure to chemicals
1) Monitoring of air
2) Ventilation
3) Personal protective equipment provided
4) Medical surveillance
5) Mechanized processes to avoid contact

iv. Personal protective measures that should be taken by the employee
1) Equipment for normal use
2) Equipment for foreseeable emergency
3) How to obtain equipment
4) Instruction on the use of equipment

NOTE TO TRAINER: *Repeat the instructions below for each chemical present in the work area. Chemicals with the same hazard properties that are used in the same way may be combined.*

5. Labeling plan description

a. Responsibilities to ensure that labels are affixed
b. Employees' responsibilities
c. What to do if an improperly labeled contained is discovered
d. What, if any, disciplinary action will be taken against offenders
e. Referring to the data sheet after reading the labels

6. Foreseeable emergencies and employee self-protection

a. Description of possible emergencies, such as fuel fires, or equipment rollovers
b. Description of measures that employees should take in various emergency situations, such as evacuation or use of personal protective equipment

7. **Summary of information and availability to employees**

    a. Package labels
    b. Data sheets
    c. Company individuals who are available to answer further questions, and provide additional information

10.7.3 Hazard Communication Training Log

The ***(COMPANY) Hazard Communication Training Log*** appearing on the following page is to be used to record attendance of employees at training programs presented by the **(COMPANY)** with respect to hazard communications and the HAZ-COM standard.

The record is to be maintained by the Safety and Health Manager, and kept current at all times.

COMPANY LOGO	Construction Safety and Loss Control Program Manual	Page: 10-11	Appr:
	10. HAZARD COMMUNICATION PROGRAM	Revision Date:	

(Sample)

(COMPANY) Hazard Communication Training Log

Employee Name (Print)	Employee Signature	Date

10.8 EMPLOYEES OF SEPARATE CONTRACTORS AND SUBCONTRACTORS

Upon notification by responsible company project management or superintendents, the company Safety and Health Manager or a Project Manager will advise outside subvendors in person of:

- Any chemical hazards that may be encountered in the normal course of their work on the respective jobsite,
- The labeling system in use for hazardous materials,
- The protective measures to be taken by all subvendor employees,
- The safe handling requirements and procedures to be used, and
- The location and availability of all Material Safety Data Sheets (MSDSs).

Each separate contractor or subcontractor bringing any chemical on-site *must* provide (COMPANY) with appropriate hazard information on all substances, including the labels used and the precautionary measures to be taken in working with these chemicals.

10.9 CURRENT LISTING OF MSDSs

NOTE TO MANUAL PREPARER: The listing of materials in the section that follows is an example of one list prepared and used by a site development company. For your own actual Safety and Loss Control Program Manual, identify your own list of chemicals and substances used or potentially used.

The listing of Material Safety Data Sheets (MSDSs) that follows is current as of the date of this Manual revision, and is provided here for convenience only.

A copy of the actual MSDSs for each product is contained in the **(COMPANY)** ***Hazard Communication Program Manual*** being maintained by the **(COMPANY)** Safety and Health Manager.

Check with the office of the Safety and Health Manager for a current listing, and if any information is needed for items not listed.

Anionic Asphalt Emulsion
Asphaltic Concrete Surface Course (hot mix)

Calcium Hypochlorite Hydrated (Chlorine)
Charlotte Pipe Seal Adhesive—Lubricant
Coal Tar Emulsion (asphalt coating)
Crushed Limestone
Diesel Fuel (No. 2)
Fire Extinguishing Agents—Dry Chemical—ABC
Fire Extinguishing Agents—Dry Chemical—"Purple K"
Fire Extinguishing Agent—Dry Chemical (sodium bicarbonate)
Fire Extinguishing Agents—Dry Chemical—Regular
Fire Extinguishing Agent—Carbon Dioxide (CO_2)
Fire Extinguishing Agent—Halon 1211 ($CFrClF_2$)
Gasoline (super unleaded)
Gear Oil (SAE 80, 90, and 140)
Grey Iron Castings
Hydrocarbon Concrete Release Agents
Instant Hydraulic Cement (high alumina)
Isopropyl Alcohol (1%)
(Manufacturer Name) Laminate Plywood
Motor Oil (HD)
(Manufacturer Name) Spray Enamel, Gloss Black 1710
(Manufacturer Name) Spray Enamel, Royal Blue 1710
(Manufacturer Name) Clear PVC Solvent Cement
(Manufacturer Name) Purple Primer NSF
(Manufacturer Name) Cleaner
Outboard Engine Oil
(Manufacturer Name) PVC to Flex (Pipe Cement)
(Manufacturer Name) All Purpose Cleaner (biodegradable, environmentally safe)
Petroleum Asphalt
Pipe—Copper
Pipe—Red Brass
Pipe Joint Lubricant
Pipe Nipples—Aluminum
Pipe Nipples—Copper
Pipe Nipples—Red Brass
Pipe Nipples—Red Brass (chrome plated)
Pipe Nipples—Standard
Portable Toilet Chemical Deodorant
Portland Cement (white and grey)
PVC—Pipe Cement
PVC—Pipe and Fittings
PVC—Primer
Red Shopcoat Primer

(Manufacturer Name) No. 5 Pipe Thread Sealer
(Manufacturer Name) CPVC Solvent Cement
(Manufacturer Name) T Plus 2
(Manufacturer Name) 105 Solvent
Silica Sand
Silicon Spray
Standard Pipe Nipples
(Manufacturer Name) Paste
(Manufacturer Name) Spray Paint
Tubing—Copper
(Manufacturer Name) Injection Polyurethane
Unhardened Concrete (redi-mix)
(Manufacturer Name) Premium Multi-Purpose EP Grease

10.10 EMERGENCY FIRST AID—PETROLEUM, CONCRETE/CEMENT, AND ASPHALT PRODUCTS

10.10.1 Overview

NOTE TO MANUAL PREPARER: Review the information below with appropriate health care professionals who can confirm the information or provide more complete/correct information with respect to the subject. Add information for all other specific situations as may be available.

The information contained in this section is very basic. It is intended to be used in providing *emergency* first aid for conditions of contact with petroleum, concrete/cement, and asphalt based materials. It should be coordinated with the First Aid Procedures more completely described in Section 4 of this ***Safety and Loss Control Program Manual***.

Although the Material Safety Data Sheets (MSDSs) have the information necessary to inform trained medical specialists in the treatment of an emergency, the situation may occur when an employee may need to quickly know first aid facts, and be able to administer care that may be essential in saving a life.

Please read the few items below very carefully. Fortunately, many people may find that this information, as well as any CPR and first aid certification training, may become surprisingly automatic in the event of a real emergency. Above all, remember not to panic in the event of any emergency. Stay calm, in order to preserve the greatest opportunity to be effective in bringing an emergency under control.

In all cases, get immediate professional medical attention as soon as the emergency measures have been taken.

As a supplement, it may be advisable to use the separate topics below as additional subjects of Superintendent's Tailgate Safety Meetings. These short, regular meetings can be an excellent forum for routinely bringing this information to the attention of all company employees.

10.10.2 Petroleum Products

Examples: Gasoline, Diesel Fuel, Kerosene, Mineral Spirits, Etc.

1. ***Never induce vomiting*** when a petroleum-based product has been ingested.
   a. Vomiting may cause aspiration of liquid into lungs, risking chemical pneumonia, as well as intestinal or central nervous system effects.
2. If swallowed, give ***water*** or ***milk*** if available.
3. Should spontaneous vomiting occur, monitor breathing.
   a. If not breathing, institute CPR. Have someone contact 911 for assistance.
4. Although petroleum asphalt is a petroleum product, ingestion is not likely, and it has a low order of acute oral toxicity. The main danger would be accumulation of hydrogen sulfide gas.
   a. If inhaled, move the person to fresh air.
   b. If the person is not breathing, administer CPR and contact 911 for assistance.

10.10.3 Concrete and Cement Products

1. Eyes
   a. Flush immediately and repeatedly with water.
2. Skin
   a. Wash exposed skin areas promptly with water

10.10.4 Asphalt Products

1. Hot asphalt causes severe burns.
   a. ***Do not attempt to remove hot asphalt.***
   b. Follow the cooling process below while awaiting emergency personnel or transportation to the hospital.
   c. Natural separation will occur within 48 to 72 hours.
   d. For more rapid removal, leave up to the physician.

2. Eyes
   a. Flush gently with large amounts of water
   b. ***Immediately call a physician.***

3. Skin
   a. Cool the hot asphalt and affected parts of the body immediately in this order of preference. ***Do not apply ice.***
      1) Completely submerge affected area in ice water.
      2) Completely submerge affected area in tap water.
      3) Place affected area under running water.

Use any available water cooler than body temperature, while trying to get to the hospital. The injured person should not drive to the hospital himself or herself.

Avoid inhalation of fumes—use in well ventilated areas.

10.11 SAMPLE MSDS LETTERS

10.11.1 General Description

The sample letters that follow are designed as a convenience to all **(COMPANY)** employees to use whenever communications regarding MSDSs and their distribution are required. They may be used by any company employee who is charged with the responsibility to secure MSDSs and/or communicate information regarding MSDSs to any project participant.

Each should be retyped on company letterhead. Copies of each completed letter are to be maintained in the appropriate project files, and distributed to the Safety and Health Manager.

10.11.2 Sample MSDS Info Letter—(COMPANY) to General Contractor or Construction Manager

The sample letter that follows below may be used in those cases where **(COMPANY)** is operating as a subcontractor or trade contractor on a project; where the prime contractor is either a general contractor (GC) or construction manager (CM). As such, **(COMPANY)** is a member of the project team, which is under the direction of the GC or CM (sample letter on next page).

***** LETTERHEAD *****

(Date) ***CERTIFIED MAIL***
RETURN RECEIPT REQUESTED

(General Contractor or Construction Manager)

RE: *(Project Name)*
(Project Number)

SUBJ: Material Safety Data Sheets and List of Hazardous Chemicals

Dear *Mr./Ms. (Project Manager)*:

Please find enclosed two copies of the complete list of hazardous chemicals that (**COMPANY**) is using or may use on the referenced jobsite. Two copies of the individual Material Safety Data Sheets (MSDSs) is also included for each of material identified.

These materials are being provided for your use and the use of all other subcontractors *(trade contractors)* on the site.

Please advise this office of the location of all such information assembled from all other project subcontractors *(trade contractors)*, in order to ensure that our field staff has access to all such information.

Thank you for your consideration.

Very truly yours,

(**COMPANY**)

cc: Safety and Health Manager

10.11.3 Sample MSDS Info Letter—(COMPANY) to Subcontractor

The sample letter that follows may be used in those cases where **(COMPANY)** is operating as the prime contractor, either as a general contractor or construction manager on a project. The letter is to be sent to subcontractors or trade contractors under the direct control of **(COMPANY)**, where **(COMPANY)** is acting as the controlling member of the project team. In this capacity, **(COMPANY)** is responsible for the assembly of all such information, and for the establishment of a consolidated file that is accessible by all project participants (sample letter on next page).

COMPANY LOGO	Construction Safety and Loss Control Program Manual	Page: 10-19	Appr:
	10. HAZARD COMMUNICATION PROGRAM	Revision Date:	

***** LETTERHEAD *****

(Date) ***CERTIFIED MAIL***
RETURN RECEIPT REQUESTED

(To each Subcontractors or Trade Contractor)

RE: *(Project Name)*
(Project Number)

SUBJ: Material Safety Data Sheets and List of Hazardous Chemicals

Dear *Mr./Ms. (Subcontractor Project Manager)*:

Please find enclosed the complete list of hazardous chemicals identified to date that (**COMPANY**) and all subcontractors *(trade contractors)* are using or may use on the referenced jobsite. This information is being provided to you as required by law.

Please immediately provide the list of all hazardous chemicals as required by your subcontract and by law that your company will use or may use on the referenced project, along with two copies of the Material Safety Data Sheets (MSDSs) for each material.

This information will be consolidated with that of all other project participants, and will be maintained in the Hazardous Chemicals and Materials File at the (**COMPANY**) jobsite field office. It will be available at all times to all project participants to review and copy as may be necessary.

Thank you for your cooperation.

Very truly yours,

(**COMPANY**)

cc: Safety and Health Manager

COMPANY LOGO	Construction Safety and Loss Control Program Manual	Page: 10-20	Appr:
	10. HAZARD COMMUNICATION PROGRAM	Revision Date:	

10.11.4 MSDS Info Letter—(COMPANY) to Fire Department

The sample letter that follows may be used in those cases where **(COMPANY)** is operating as the prime contractor; either as a general contractor or construction manager on a project. The letter is intended as an informational notice to be sent to the local fire department station that is nearest to the jobsite. It may be combined with a more complete letter of introduction, or may be used by itself if a more regular line of communication has already been established between the project staff and the local fire department personnel (sample letter on next page).

***** LETTERHEAD *****

(Date)

(To fire department station nearest to the jobsite)

RE: *(Project Name)*
(Project Number)

SUBJ: Material Safety Data Sheets and List of Hazardous Chemicals

Dear Captain *(Subcontractor Project Manager)*:

Please find enclosed the complete list of hazardous chemicals identified to date that (**COMPANY**) and all subcontractors *(trade contractors*) are using or may use on the referenced jobsite.

This material is being provided to you for informational purposes.

Please feel free to contact me at (telephone number) or at the above address at any time if you require any additional information.

Thank you for your consideration.

Very truly yours,

(**COMPANY**)

cc: Safety and Health Manager

10.11.5 MDSD Request Letter

The sample letter that follows may be used in the effort by (**COMPANY**) to secure Material Safety Data Sheets (MSDSs) from chemical or other product manufacturers who have either neglected to provide the MSDSs with their product shipments, or to secure the respective MSDSs prior to any particular product shipment (sample letter on next page).

***** LETTERHEAD *****

(Date) ***CERTIFIED MAIL***
RETURN RECEIPT REQUESTED

(To each product manufacturer of a chemical or other substance to be used on the project)

RE: *(Project Name)*
(Project Number)

SUBJ: Material Safety Data Sheets
(List Products)

Dear *Mr./Ms. (Individual)*:

Please forward to my attention the Material Safety Data Sheet (MSDS) for each of the subject products identified above which are intended to be used on the referenced project. Please be sure that the information is in form and content as required by the Federal Hazard Communication Standard.

If any of the *subject* products are not *subject* to the Federal Hazard Communication Standard and are therefore not required to have an MSDS on file at the project site, please provide your written confirmation of this fact.

Thank you for your consideration.

Very truly yours,

(COMPANY)

cc: Safety and Health Manager

10.12 ADDITIONAL INFORMATION

All (**COMPANY**) employees or their designated representatives can obtain further information on this written program, the Hazard Communication Standard, applicable MSDSs, and chemical information lists at the office of the Safety and Health Manager.

In emergency situations, always call a 911 and/or doctor as appropriate, as quickly as possible. Use your judgment as to the necessity of any emergency procedure.

Sample Section 11

GENERAL EMERGENCY PROCEDURES

Section Use

The form and content of Sample Section 11 *General Emergency Procedures* is designed as an example of company instruction to its employees with respect to the material presented. It is intended to provide for each employee:

- Guidance in getting emergency situations under control as expeditiously as possible, and effectively resolving immediate emergency situations.
- Direction in the most expedient manner to involve all appropriate company officials and staff.
- A summary of emergency procedures that can be referred to by all company employees as needed.

The sample section that follows is an example of a company safety and loss control program approach to the objectives indicated above. The sample language and forms are to be considered in the context of your own company and its manner of operation.

Carefully consider these objectives during the review and editing process of the section material in order to customize the section content to the specific needs of your organization. Prepare final language that is consistent with decisions made by your own management in the determination of your own company program.

It is important to understand that safety rules and regulations change from time to time. Be sure to review the final draft of the section and the entire program manual with appropriate safety, insurance, and legal professionals prior to issue and use. Periodically review your program with your own company management and all appropriate safety, insurance, and legal professionals to be certain that your own company program remains current with all requirements.

COMPANY LOGO	Construction Safety and Loss Control Program Manual	Page: 11-1	Appr:
	11. GENERAL EMERGENCY PROCEDURES	Revision Date:	

(Sample)

SECTION 11

General Emergency Procedures

COMPANY LOGO	Construction Safety and Loss Control Program Manual	Page: 11-2	Appr:
	11. GENERAL EMERGENCY PROCEDURES	Revision Date:	

11.1 INTRODUCTION

NOTE TO MANUAL PREPARER: The following sections are representative of recommendations that have been made by various insurance and health care providers for use by contractors, and are examples of a company's method of disseminating that information to its employees. Check with your own company's insurance and health care providers to see if they can provide versions of this type of material that may be more appropriate for your use in your own actual Safety and Loss Control Program Manual.

It is the company's intention to prepare and update relevant procedures for governing actions to be taken in the event of serious injury, property damage, or other catastrophe as company operations will require. Such emergency procedures may include:

- Necessary action to be taken,
- Who should be responsible to take those actions,
- The names of all individuals and/or entities that should be notified, and
- The location(s) of emergency equipment and supplies.

All supervisory and other key (**COMPANY**) personnel should be thoroughly familiar with these procedures.

Copies of these procedures are to be made available to every (**COMPANY**) employee upon request, and can be obtained directly from the office of the (**COMPANY**) Safety and Health Manager.

Note that this section is for general direction. All instruction must be coordinated and performed in accordance Section 6 of this ***Safety and Loss Control Program Manual*** for following procedures regarding ***Accident Investigation, Reporting, and Records***.

11.2 GENERAL

Copies of the emergency procedures should be given to all (**COMPANY**) contractors and subcontractors with their original subcontract. The Site Superintendent should review the program with each contractor or subcontractor in order to be certain that those parties understand the requirements and their respective responsibilities.

The Site Superintendent shall note in his or her Daily Field Report, in any minutes of Safety Meetings, or in other relevant project documentation that the procedure was reviewed with each party on the site. Securing their signature on the copy of the respective Safety Meeting form will also be sufficient documentation.

11.3 GENERAL EMERGENCY PROCEDURES

Where specific procedures may not have been established, judgment shall be used in determining the best course of action. Follow the guidelines below for assistance:

- All emergencies are to be handled by the highest ranking person present, to be assisted by whomever may be available.
- The ranking person shall delegate responsibility for making emergency phone calls.
- Emergency phone numbers are to be placed in conspicuous places throughout the jobsite.
- The need for an ambulance or other emergency equipment shall be determined by the ranking person present, or by any medical personnel present, except where a catastrophic event has occurred. In the event of a catastrophic occurrence, any one person on the site may call for an ambulance.
- In all instances, the **(COMPANY)** home office shall be notified immediately by phone, upon completion of any necessary emergency first aid treatment.
- General fire protection procedures described in the following Section 11.4 apply to all projects, including those with vertical as well as horizontal construction. Site development projects with minimal fire risk (open road projects, for example) may not require the complete attention as indicated. In such cases, the Project Manager in conjunction with the **(COMPANY)** Safety and Health Manager will at the onset of the project determine the complete program appropriate for the specific project.

11.3.1 Emergency Procedure Summary Form

The Emergency Procedure Summary Form on the following page is a reiteration of the previous Section 11.3. It is to be used to comply with the requirements to post such procedures as described above, and to use as a handout to subcontractors on the jobsite.

Contact the (**COMPANY**) Safety and Health Manager if there are any questions on the use of the form.

COMPANY LOGO	Construction Safety and Loss Control Program Manual	Page: 11-5	Appr:
	11. GENERAL EMERGENCY PROCEDURES	Revision Date:	

(Sample)

Emergency Procedure

- All emergencies are to be handled by the ranking person present, with whomever is available to assist.
- The ranking person shall delegate responsibility for making emergency phone calls.
- Emergency phone numbers are to be placed in conspicuous places throughout the jobsite.
- The need for an ambulance or other emergency equipment shall be determined by the ranking person present, or by any medical personnel present, except where a catastrophic event has occurred. In the event of a catastrophic occurrence, any one person on-site may call for an ambulance.
- Where specific procedures have not been established, relative judgment shall be used in determining course of action.
- In all instances, the (**COMPANY**) home office shall be notified immediately by phone or radio, after any necessary emergency first aid treatment has been administered.
- General Fire Protection procedures described below apply to all projects, including those with vertical as well as horizontal construction. Site development projects with minimal fire risk (open road projects, for example) may not require the complete attention as indicated. In such cases, the Project Manger in conjunction with the (**COMPANY**) Safety & Health Manager will at the onset of the project determine the complete program appropriate for the specific project. Considerations for the complete program will include the following:
  - The Project Manager and the (**COMPANY**) Safety & Health Manager will make regular site inspections in attempt to identify fire hazards, and to make recommendations for correction.
  - Prior to job start-up, each (**COMPANY**) subcontractor will attend a fire safety planning conference conducted by (**COMPANY**). The minutes of the conference shall include a list of all fire hazards anticipated during the work, along with recommendations for prevention.
  - All flammable liquids will be protected with appropriate water or dry powder extinguishers as required.
  - Fire alarm instructions will be posted throughout the site with telephone numbers for local fire departments.
  - Each (**COMPANY**) subcontractor will identify two (2) competent employees who will be available to work on fire safety problems upon request. A list of names of these individuals with telephone numbers will be posted at the(**COMPANY**)field office, if there is one, and in any event will be maintained on file by the (**COMPANY**) Project Manager.
  - Each worker will be thoroughly informed of any special fire hazards to be anticipated during the work of the procedures to be used to avoid starting a fire. Each (**COMPANY**) subcontractor will be required to keep a record to indicate when and to whom this instruction was given on forms available from the Project Manager.
  - Extinguishers shall be placed on numbered stations with a corresponding number affixed to the extinguisher. Extinguishers are to be checked and inspected to be in ready-use conditions at least weekly, with a record of this inspection made and posted.
  - In the interest of fire safety, (**COMPANY**) will control the storage of flammable materials to avoid large concentrations of such materials at one place.
  - Hydrants, standpipes, and sprinkler systems will be activated as soon as practical after installation, in order to be available as aids in suppressing fire.
  - Any guard forces implemented on the site will be instructed to make special inspections after working hours to assure that fires, if inadvertently started during work, or started for a purpose and then extinguished, are not continuing. Guard forces will be instructed in the use of fire first-response equipment and will be thoroughly familiar with plans for summoning local units to fight fire after working hours, on Saturdays, Sundays, and holidays.

11.4 FIRE PROCEDURE

General fire protection procedures described in this section apply to all projects, including those of both vertical (building) as well as horizontal (road and sitework) construction. Site development projects with minimal fire risk (open road projects, for example) may not require the complete attention as indicated. In such cases, the Project Manager in conjunction with the (**COMPANY**) Safety and Health Manager will at the onset of the project determine the complete program appropriate for the specific project.

Considerations for the complete Fire Protection Program will include the following:

- The Project Manager and the (**COMPANY**) Safety and Health Manager will make regular site inspections in an attempt to identify fire hazards, and to make recommendations for correction.
- Prior to job start-up, each (**COMPANY**) subcontractor will attend a fire safety planning conference conducted by (**COMPANY**). The minutes of the conference shall include a list of all fire hazards anticipated during the work, along with recommendations for prevention.
- All flammable liquids will be protected with appropriate water or dry powder extinguishers as required.
- Fire alarm instructions will be posted throughout the site with telephone numbers for local fire departments.
- Each (**COMPANY**) subcontractor will identify two (2) competent employees who will be available to work on fire safety problems upon request. A list of names of these individuals with telephone numbers will be maintained on file by the (**COMPANY**) Project Manager.
- Each worker will be thoroughly informed of any special fire hazards to be anticipated during the work of the procedures to be used to avoid starting a fire. Each (**COMPANY**) subcontractor will be required to keep a record to indicate when and to whom this instruction was given.
- Extinguishers shall be placed in appropriate locations and regularly inspected to be in ready-use condition.
- In the interest of fire safety, (**COMPANY**) will control the storage of flammable materials to avoid large concentrations of such materials at one place.
- Hydrants, standpipes, and sprinkler systems will be activated as soon as practical after installation, in order to be available as aids in suppressing fire.

- Any guard forces implemented on the site will be instructed to make special inspections after working hours to assure that fires, if inadvertently started during work, or started for a purpose and then extinguished, are not continuing. Guard forces will be instructed in the use of fire first-response equipment and will be thoroughly familiar with plans for summoning local units to fight fire after working hours, on Saturdays, Sundays, and holidays.

In summary, emergency procedures to be used in the event of a fire on a jobsite or at any company facility include the following:

- Make a safe attempt to extinguish. Do not endanger your life.
- At the same time, notify the Fire Department.
- Have an employee meet the Fire Department at the site entrance to direct them to the location of the fire.
- Keep all spectators and non-essential employees away from the fire.
- If explosive-type materials are involved, immediately evacuate all personnel.
- Make no comments. Refer all inquiries to the (**COMPANY**) Project Manager.

11.5 ACCIDENTS INVOLVING SERIOUS INJURY OR DEATH

- Provide for necessary first aid.
- ***Call 911***, or have first available person call 911, and advise of complete situation.
- Notify the Site Superintendent, Project Manager, and the Safety and Health Manager.
- Remove and/or keep back all non-essential personnel.
- Provide assistance to rescue personnel as requested.
- Make no comments. Refer all inquiries to the (**COMPANY**) Project Manager.
- Allow no on-site pictures to be taken, except upon approval of the (**COMPANY**) Project Manager or Safety and Health Manager.
- Make full investigation, and report per the requirements of Section 6 of this ***Safety and Loss Control Program Manual***.

COMPANY LOGO	Construction Safety and Loss Control Program Manual	Page: 11-8	Appr:
	11. GENERAL EMERGENCY PROCEDURES	Revision Date:	

11.6 PROPERTY DAMAGE ACCIDENTS

- Notify the Site Superintendent or Project Manager at first availability. The Superintendent or Project Manager should immediately notify or arrange notification to the **(COMPANY)** Safety and Health Manager.
- Protect against further damage whenever possible.
- Where the possibility of fire, electrical injury, or explosion exists, take additional measures as necessary to protect.
- Keep all spectators and non-essential employees back and/or away from the area.
- Make no comments. Refer all inquiries to the **(COMPANY)** Project Manager or Safety and Health Manager.
- To the extent possible, allow no on-site pictures to be taken, except upon approval of the **(COMPANY)** Project Manager or Safety and Health Manager.
- Make full investigation, and report per the requirements of Section 6 of this *Safety and Loss Control Program Manual*.

11.7 PUBLIC DEMONSTRATIONS

- Notify the **(COMPANY)** Project Manager and Safety and Health Manager, who will notify appropriate authorities.
- Remove all **(COMPANY)** employees from the demonstration area.
- Do not allow any verbal or physical confrontations, except for personal protection.
- If the conditions becomes worse and no **(COMPANY)** supervisory personnel are present, **(COMPANY)** employees are to report to the site office and await further instruction.

11.8 BOMB THREATS

When a bomb threat is received for the first time:

- The project or office shall be evacuated immediately.

COMPANY LOGO	Construction Safety and Loss Control Program Manual	Page: 11-9	Appr:
	11. GENERAL EMERGENCY PROCEDURES	Revision Date:	

- Evacuation will be made of all personnel on the project or in the office.
- A count should be made to assure that all are present and accounted for.

- Local police, fire or bomb disposal authorities will be immediately notified.
- Do not allow anyone except authorized personnel to reenter the area.
- If necessary to stop or detour traffic away from the affected a area, utilize local police or flagmen.
- A search of the premises will be made by the appropriate authorities. No (**COMPANY**) supervisory personnel or other employees are to be involved with the search.
- Notify any affected businesses or residents who may be endangered.
- Appoint one person (generally the ranking person present) as the (**COMPANY**) spokesperson.
- Allow no photographs on site.
- Make no comments.
- Notify the (**COMPANY**) home office immediately.
- Assist the authorities with information and in every other way possible, without placing any (**COMPANY**) employee in any danger.
- If a suspicious article is found—***do not touch it.*** Have the appropriate authorities handle the situation from that point.

If no bomb is found and a second threat is made shortly after the first one, the premises should be evacuated again and searched again by the appropriate authorities. If a third threat occurs under circumstances which make the threat appear to be a continuation of the same pattern of false claims, careful evaluation of the circumstances is to be made by the company senior management in consultation with appropriate law enforcement authorities.

If a threat occurs after a substantial period of time has elapsed since the previous threat, or if for any reason the threat seems to be unrelated to the earlier threats, the above procedures should be followed as indicated for the initial threat.

Sample Section 12

SUBCONTRACTOR JOBSITE SAFETY POLICY

Section Use

The form and content of Sample Section 12 *Subcontractor Jobsite Safety Policy* is designed as an example of company instruction to its employees with respect to the material presented. It is intended for use by each employee as appropriate:

- To provide an actual example of a company safety policy and program that is to used as an subvendor as an attachment to all company subcontracts. It itemizes the subcontract requirements with respect to safety, and highlights work activities that will need regular attention
- To communicate the safety and loss control program, directly incorporate those requirements into the respective agreements, and communicate to each subcontractor the expectation of their compliance with the information as it is presented.

The sample section and program that follows is an example of a company safety and loss control program approach to the objectives indicated above. The sample language and the example subcontract attachment itself is to be applied in the context of your own company and its manner of operation. It is an important legal document that should be carefully reviewed by your own attorney and safety consultant prior to implementation in any form.

Carefully consider these objectives during the review and editing process of the sample section material in order to customize the section content to the specific needs of your organization. Prepare final language that is consistent with decisions made by your own management in the determination of your own company program.

It is important to understand that safety rules and regulations change from time to time. Be sure to review the final draft of the section and the entire program manual with appropriate safety, insurance, and legal professionals prior to issue and use. Periodically review your program with company management and all appropriate professionals to be certain that the company program remains current with all requirements.

COMPANY LOGO	Construction Safety and Loss Control Program Manual	Page: 12-1	Appr:
	12. SUBCONTRACTOR JOBSITE SAFETY POLICY	Revision Date:	

(Sample)

SECTION 12

Subcontractor Jobsite Safety Policy

12.1 SUBCONTRACTOR JOBSITE SAFETY POLICY

12.1.1 Description

The Subcontractor Jobsite Safety Policy included in this section is intended to summarize the general requirements included in each **(COMPANY)** subcontract and trade contract, and present them in a consolidated package. The package will:

- Clarify (**COMPANY**) safety policy,
- Identify requirements and expectations of each subcontractor with respect to safety.

12.1.2 Section Use

NOTE TO MANUAL PREPARER: The section 12.1.2 material below is self explanatory. It is critical, however, that your own company management clearly understand the intention of the material to be routinely reproduced as a specific attachment to each of your company's subcontracts. It is therefore imperative that the final subcontract package ultimately put into practice by your own company meet the objectives stated in this subsection. Review it carefully, and coordinate it closely with the preparation of your own company standard subcontract form.

With the exception of the introductory material and explanations of Section 12.1, the entire Section 12 is intended to be photocopied in its entirety, with the specific project identification information completed, and used essentially as-is as an attachment to each subcontract or trade contract for each respective project.

The **(COMPANY)** project manager is to review the materials herein, specifically verify that all requirements of the **(COMPANY)**/Client Agreement are accommodated, and make any changes or corrections that may be necessary to ensure that the subcontract package meets all specific project requirements prior to final issue for use.

12.2 SAMPLE ATTACHMENT TO SUBCONTRACTS—SUBCONTRACTOR JOBSITE SAFETY PROGRAM

(Sample)

Attachment ___
To the Subcontract dated

____

by and between
(COMPANY)
and

(*Subcontractor Name:*) ________________________ (Subcontractor)

(COMPANY)
Subcontractor Safety Program

General Guidelines

This (**COMPANY**) Subcontractor Safety Program is provided in order to inform the subcontractor of the (**COMPANY**) policy with respect to safety and health on all (**COMPANY**) jobsites, and to amplify certain details of the complete requirements of the subcontract agreement. It is not intended to abridge or restrict any safety requirement of the subcontract, but is intended as a supplement.

Specific rules, conditions, forms, and requirements have been cataloged in this Subcontractor Safety Program, which shall govern the conduct of the subcontractor and all subcontractor employees while working in and around any (**COMPANY**) jobsite. The rules an practices set forth in this program are those with broad application. It will often be necessary to supplement the information in this program with additional rules and information specific to the individual project. To that extent, it is subcontractor's responsibility to be aware of all such information, and to comply with all requirements in every respect.

This material has been prepared with the intention of highlighting compliance with current Occupational Safety and Health Administration standards and regulations, and with all current laws of the United States. If any error, omission, or conflict should arise between the material of the subcontract and/or the material presented herein with any applicable standards, regulations, or laws, the laws and regulations governing the jobsite location shall prevail.

Safety Policy

It is the Policy of (**COMPANY**) to at all times:

- Provide safe working conditions,
- Perform all activities in ways that eliminate risk of injury or health impairment to any tradesperson,
- Maintain all areas in ways that eliminate risk to visitors and to the public, and
- Eliminate risk of damage to property on and adjacent to every jobsite.

These are basic responsibilities of every company and individual on every jobsite. All supervisors of all trades must routinely accept complete responsibility for prevention of accidents and for the safety of all work under their direction. All tradespeople of every category are required to conduct themselves in a safe, considerate, and workmanlike manner.

By contract and by law, every company and person employed on the site is obligated at a minimum to comply with the Federal Occupational Safety and Health Act, Americans with Disabilities Act, and the laws of every entity having jurisdiction over the work and the site.

Any company or individual refusing to correct observed safety violations will be banned from the site at least until such violations are corrected, and will be held completely responsible for all resulting effects.

With attention, awareness, and cooperation by everyone, we will all achieve an accident-free job, and the pride that goes with it.

Subcontract Requirements

NOTE TO MANUAL PREPARER: To the extent that these or similar terms are included directly in the body of your company's standard subcontract form, the respective requirements may either be deleted from this package and only be referred to herein. Such requirements, however, may still be repeated here for emphasis. All text that follows is a critically important legal document that must be reviewed and approved for use by your attorney prior to implementation in any way.

By executing the subcontract, the subcontractor agrees that the work is to be completed in accordance with all applicable codes and government and quasi-government agencies having jurisdiction over the Work of the subcontract. To the extent of the work performed by the subcontractor, the subcontractor shall be bound to (**COMPANY**) and shall assume toward (**COMPANY**) all the obligations and responsibilities which **(COMPANY)** by its Contract Documents assumes toward the Owner.

Immediately upon execution of the subcontract, subcontractor shall provide a list of all hazardous chemicals and material that will or may be used in the prosecution of its work on the site, will provide Material Safety Data Sheets (MSDSs) for all materials prior to their delivery, and will maintain an additional copy of all this information on the jobsite available for inspection by any project contractor or subcontractor at all times during regular business hours. All submissions relating to the subcontractor's Hazard Material Management Program are to be submitted to **(COMPANY)** prior to the start of any work on the jobsite.

Upon delivery, the subcontractor is to immediately verify that all materials comply with all related safety and health information as delivered, and with all approved submittals. The subcontractor shall strictly follow material and equipment manufacturer's latest printed instructions for delivery, handling, storage, protection, installation, execution, and cleaning.

All work must be done by skilled labor in accordance with the highest quality levels and most recent standards of industry practice, applicable building codes, OSHA Safety and Health Standards and Regulations, Worker's Compensation Laws, Drug Free Work Place Laws, the OSHA Trench Safety Act, and all other requirements of the Contract Documents, and by any government or quasi-government agency having jurisdiction over the Work. No radios, tape players, signs, or advertisements of any kind are allowed on or in the vicinity of the jobsite.

The subcontractor shall conduct weekly Safety Meetings with its work force on-site for a minimum of five (5) minutes in length with printed handout materials for each meeting participant, will record minutes of each meeting (including list of attendees), and will forward a copy of the minutes and handouts to **(COMPANY)** on a weekly basis.

As applicable to its work, the subcontractor shall backfill and compact excavations daily and/or whenever possible; ongoing excavations not practical to be backfilled daily shall be properly barricaded and protected.

The subcontractor shall: (a) not park equipment on roadways, near ponds, gates, wetlands, sink holes, over or blocking items requiring access, or in any manner that might risk physical or environmental damage or interfere with site access; (b) not leave any equipment unattended with engines running; and (c) be responsible for any fuel, oil, liquid, or other spills, including their legal cleanup and disposal.

Subcontractor shall provide safe and sufficient facilities at all times for inspection of the Work by **(COMPANY)**, the Owner, or design professionals and/or their representatives, and shall provide all barricades, fencing, signs, protections, and temporary constructions necessary to prevent the public and other unauthorized personnel from entering the site or proximity of the Work. The subcontractor agrees that prevention of accidents to workmen engaged in the work and protection of the public throughout areas adjacent to the Work are the exclusive responsibility of the subcontractor, and shall take every measure and precaution as required by **(COMPANY)**, the Subcontract Documents, the Contract Documents, the Federal Occupational Safety and Health Act, and every other entity having jurisdiction over the Work. The subcontractor further agrees that safety, security for and protection of its materials, equipment, and the Work of its subcontract is its exclusive responsibility and highest priority until final completion and acceptance of the project by the Owner.

If the subcontractor encounters material on the site reasonably believed to be asbestos, polychlorinated biphenyl (PCB), or any other toxic or hazardous material which has not been rendered harmless, the subcontractor shall immediately stop work in the area(s) affected, and report the condition to **(COMPANY)** immediately and in writing. The work in the affected area(s) shall immediately resume when it has been determined that the suspect material is not asbestos, polychlorinated biphenyl (PCB), or any other toxic or hazardous substance, or when the material has been removed and/or rendered harmless in accordance with proper legal procedure.

Prior to any work being performed in any Confined Space as defined by OSHA, the subcontractor shall test the area for radon, oxygen level, and any air-quality hazards, and take all appropriate measures to establish and maintain required ventilation and air quality maintenance.

To the extent applicable to its work, the subcontractor accepts all responsibility for traffic control, barricading, road damage, road user fees, stabilization of surfaces for ingress and egress, control of dust and siltation, all cleanup and housekeeping, and any fines assessed against any party due to the subcontractor's failure to perform or comply. The subcontractor is responsible for the daily cleanup of its work areas, and removal of all dirt, debris, and excess materials from the jobsite on a daily basis.

In the event that the subcontractor or any or its employees or its sub-subcontractors fail to comply with any health and safety requirements, or if **(COMPANY)** deems any part of the work performed by the subcontractor is unsafe, **(COMPANY)** may require the subcontractor to stop work and/or remove any noncomplying employees, and immediately correct all noncompliances. In such event, the subcontractor shall not be entitled to any additional time or compensation of any kind as result of such stoppage, but will remain responsible to **(COMPANY)** for all costs associated with any related delay to the project by the terms of the Contract Documents.

Accident Investigation and Reporting

All accidents and occupational illnesses must be reported.

Proper investigation and reporting of accidents is necessary for all contractors and subcontractors. Determining the cause(s) of accidents will identify the steps necessary to prevent their occurrence.

It is the policy of **(COMPANY)** that every accident or injury of any kind and at any level of severity *must* be immediately reported by the employee to the Crew

Foreman, Site Superintendent, or Project Manager—even if the employee believes the incident to be minor. All injuries or accidents must be investigated immediately by the injured person's immediate supervisor, the cause determined, and the problem corrected.

In the case of serious injury, such as fractures, cuts requiring sutures, multiple injuries, unconsciousness, vision impairment, fatality, poison ingestion, respiratory distress, etc., the **(COMPANY)** Home Office and the Safety and Health Manager must be notified as soon as practical—immediately *after appropriate emergency measures have been taken.*

For every injury, the subcontractor shall provide to the **(COMPANY)** project manager a copy of the completed ***Accident Investigation Report, Employer's First Report of Injury or Illness,*** or other similar reports required by federal, state, county, and municipal, or local safety laws for all job-related injuries and illness to their employees.

All recording requirements will be initiated and preserved in accordance with OSHA requirements. Work-related accidents that must be recorded are as follows:

1. All occupational deaths, regardless of the time between injury and death, or length of illness.
2. Accidents resulting in lost workdays (either days away from work or days of restricted work activity).
3. Accidents requiring medical treatment other then first aid.
4. Accidents resulting in loss of consciousness.
5. Accidents resulting in restriction of work or inability to perform work caused by an accident.
6. Transfer temporarily or permanently to another job because of an accident.
7. Accident resulting in termination of injured or ill person.

If an on-the-job accident occurs that results in the death of an employee or in the hospitalization of three (3) or more employees, all employers on the project site, regardless of the number of employees, are required to report the accident, in detail to the following:

- To OSHA's toll-free number: **1–800–321–OSHA**, or
- To the nearest OSHA office.

These reports must be made within eight (8) hours of the incident. This requirement applies to each such fatality or hospitalization of three (3) or more employees which occurs within thirty (30) days of an incident. The only exception to this requirement is if the employer does not learn of a reportable incident at the time it occurs and the incident would otherwise be reportable, the employer shall make the report within eight (8) hours of the time the incident is reported to any agent or employee of the employer.

First Aid

The subcontractor is responsible for providing first aid and medical treatment for their own employees. The subcontractor is also responsible to ensure that the names, addresses and telephone numbers of the subcontractor's doctors, hospital, and ambulance services are conspicuously posted as required by law.

(**COMPANY**) will provide a first aid kit for use by all parties, located at its jobsite field office. The subcontractor is required to provide its own first aid kit conspicuously located in the vicinity of each of its work areas, and readily accessible at all times. Each first aid kit is to be of an appropriate size for the respective crew.

Subcontractor Employee Orientation and Training

The subcontractor shall provide and enforce an adequate ongoing safety program for the benefit of its employees. At a minimum, the subcontractor is required to:

1. Present its safety and loss control orientation program to each new employee prior to that employee's start of work.
2. Inform their employees of all safety and health rules pertaining to their particular work assignment.
3. Inform their employees of the location(s) and uses of all safety equipment and devices; such as first aid kits, fire extinguishers, personal protective devises, personal transport devices, communication equipment, etc.
4. Conduct monthly safety meetings for its supervisory employees and weekly tailgate safety meetings for all employees, including appropriate documentation of all meetings.
5. Implement a regular system of inspection of all work areas with the intention to detect and correct hazardous and potentially hazardous conditions, violations of any safety rule, and unsafe working practices.

Subcontractor Employee Conduct

All subcontractor's employees are to be made aware of the following minimum rules of conduct, and will be required to comply with all such rules. Failure to comply may result in those employees being temporarily or permanently barred from the site, at the sole discretion of (**COMPANY**).

1. Alcoholic beverages and illegal drugs are strictly prohibited.
2. Employees entering the jobsite in the possession of or under the influence of

alcohol or illegal drugs or controlled substances shall be subject to immediate ejection from the jobsite.

3. No firearms or weapons of any kind are allowed on the jobsite.
4. Fighting, gambling, stealing, soliciting, and horseplay of any kind is strictly prohibited.
5. Abusive language or disrespectful behavior is prohibited.
6. All accidents are to be reported on the same day as the accident occurrence.
7. All non-emergency treatment of accidents is to be authorized by the injured employee's immediate supervisor.
8. All employees are to be made aware of any jobsite alarms and emergency code signals.
9. Hard hats and construction grade shoes or boots are to be worn at all times.
10. Seat belts are to be worn at all times when in company vehicles and equipment.
11. Jobsite roadways and walkways are not to be blocked without prior approval of (**COMPANY**).
12. Proper hygiene will be expected of each employee.
13. All other written and spoken safety rules are to be followed explicitly.

Employee Clothing and Personal Protective Equipment

The subcontractor is responsible to ensure that all subcontractor's employees comply with minimum requirements for clothing worn in work areas, and that all subcontractor's employees have available to them and use all personal protective equipment required by their individual work assignments. Failure to comply may result in suspension of the work being performed by those employees until the clothing or equipment need is corrected. At a minimum:

1. The subcontractor is to provide and require the use of all protective devices and personal protective equipment by its employees at all times as required by their respective work activities.
2. Approved eye and face protection must be worn when conditions require. Safety glasses are required in all circumstances where there is the possibility of exposure to flying debris or particles. Side shields should also be worn whenever possible.
3. Plastic face shields should be worn wherever there is the possibility of flying particles and spraying of liquids or corrosive substances.
4. A hard hat is to be worn at all times.
5. Only full-covered leather work shoes are allowed. Sneakers, canvas shoes, or shoes that are open in any way are not allowed.
6. Shirts must be worn at all times. Sleeveless shirts and tank tops are not allowed.
7. Shorts are not allowed. Full-length pants must be worn at all times.

COMPANY LOGO	Construction Safety and Loss Control Program Manual	Page: 12-10	Appr:
	12. SUBCONTRACTOR JOBSITE SAFETY POLICY	Revision Date:	

8. Jewelry is not to be worn on the jobsite at any time. A watch may be worn unless the employee is performing any task which may result in the watch band being caught, or an object becoming lodged between the band and skin. Watch bands should be of the expansion type, so that they would slip off if they get caught.

Housekeeping and Cleanup

The subcontractor is responsible for daily cleanup of its work areas and removal of all debris from the jobsite. As part of this effort:

1. The subcontractor is to maintain good housekeeping standards at all times.
2. All work areas must be kept free of any dirt, debris, unnecessary materials, etc. to the complete satisfaction of **(COMPANY)** project management.
3. All materials placed in piles shall be done so in a manner that maintains safe clearances, and that prevents topping or movement.
4. All loose materials must be secured so they cannot be blown around or inadvertently pushed off their stacks.
5. Accumulation of any materials that may create a fire hazard is not allowed.
6. Excessive amounts of materials are not be stored on floor, roof, or other supported areas in any manner that may possibly overweight any structural area.

Tools and Equipment

The subcontractor is responsible to ensure the issue of proper tools and equipment necessary to safely perform all operations required by the subcontract. At a minimum:

1. All owned, leased, and rented tools and equipment are to be maintained in safe mechanical working condition as specifically required by the tool and equipment manufacturer.
2. Any tools and equipment that have had their guards or other safety measures removed or modified will not be allowed.
3. All damaged or worn tools shall immediately be taken out of service and properly repaired by qualified personnel. Temporary and makeshift repairs are not permitted.

Barricades and Signs

The subcontractor is responsible to barricade all areas for overhead work, protection of workers and the public from hazardous operations and hazardous areas, moving equipment, and material storage. At a minimum:

1. All barricades in or in the proximity of roadways or walkways are to have flashing lights when used dusk to dawn.
2. When work requiring barricades is completed, the barricades are to be removed from the jobsite.
3. The subcontractor must not remove any signs posted without approval from **(COMPANY)** project management.
4. All rigging for all equipment, such as ropes, chains, cables, slings, etc. is to be inspected prior to use, and as required by the specific work assignment.
5. Ground fault detectors or an assured equipment grounding conductor program are to be implemented on all jobsites as required by law.

Ladders

The subcontractor is responsible to provide ladders of sound integrity for use by its workers. In addition to all legal requirements:

1. Ladders with split or cracked side rails or damaged rungs are not allowed on the jobsite. If any such ladders are observed, they must be taken out of service until properly repaired.
2. Ladders in doorways, walkways, and other congested areas are to be barricaded or guarded.
3. All ladders are to be of adequate length, and must extend a minimum of three feet above the upper landing.
4. Ladders must be placed on solid, level footing. For straight ladders, the foot of the ladder must be placed a minimum distance from the wall equal to 1/4 the length of the ladder from the resting point.
5. All portable straight ladders are to have non-skid feet, and are to be securely tied off when placed into position.
6. Workers are to always face the ladder and use both hands when ascending or descending.
7. Workers are not to climb or descend on ladders with tools, equipment, or materials in hand.
8. Metal ladders are not to be used in the vicinity of electrical services or electrical lines.

Fire Protection

Fire protection equipment is provided throughout the jobsite by **(COMPANY)** as required for general work conditions. The subcontractor may be required to provide for adequate fire protection for its own work operations per the conditions of its subcontract. As part of this effort:

COMPANY LOGO	Construction Safety and Loss Control Program Manual	Page: 12-12	Appr:
	12. SUBCONTRACTOR JOBSITE SAFETY POLICY	Revision Date:	

1. Fire extinguishers are to be located at a maximum of 100′ to the work area.
2. Extinguishers that are in poor condition, not of appropriate size, or have been discharged are to be immediately replaced with proper extinguishers.
3. Smoking may be prohibited on the jobsite or in certain areas of the jobsite for various reasons. Under such conditions, all employees will be expected to comply with the "no smoking" requirements.
4. All flammable liquids are to be stored in approved safe containers. Contents of such containers shall be described in proper labels and identified as flammable.

Concern for the Public

The public is not allowed free access to jobsite areas. In all situations, work activity must be performed in manners that avoid the possibility of exposure of the public to any jobsite hazards. In order to further this objective:

1. Unauthorized persons are not permitted access to any construction area.
2. Barricades and warning signs are to be posted in areas that are conspicuous to the public.
3. Be considerate of the locations of intake air ducts, air conditioners, air pumps, and similar constructions where work is performed at or near occupied areas; avoid allowing dust or vapors to be drawn into them.
4. At the end of each work day, be sure to lock all trucks and equipment, deactivate hoisting equipment, and remove all ladders if there is any possibility of access to them by the public.

Conclusion

It should be clear that all efforts taken to promote and preserve safe working environments for all workers and the public should be considered as planning and preparation time well spent. Such planning and preventative precautions will improve all operations on an ongoing basis. The most efficient use will be made of available labor, materials, and equipment, specifically resulting from reduced down time and work interruption.

Accidents are too often the result of improper planning, bad judgment, or the lack of respect for the need for proper attention in all places, at all times, and under all conditions. If the project team preserves its total commitment to themselves, to each other, to all workers, and to the public, we will preserve the best possibility for a safe, accident free, and productive jobsite, and the pride and satisfaction that will accompany it.

Sample Section 13

(COMPANY) New Hire Personal Safety Program

Section Use

The form and content of Sample Section 13 *(***COMPANY***) New Hire Personal Safety Program* is designed as an example of company instruction to its employees with respect to the material presented. It is intended to provide for each employee involved in the human resources area a condensed version of the company Safety and Loss Control Program that is to be included either as a separate booklet, or as a section in a company employee booklet that describes all procedures necessary to communicate the information to new hires. It is intended to summarize the complete program, to instruct new employees as to the content and location of the complete program, and to clearly notify new employees of the company's expectation of their compliance with the program.

The manual preparer must coordinate this section material and its objectives very closely with the company human resources department, and any effort by that department related to the preparation of a complete "new employee manual."

The sample section that follows is an example of a company safety and loss control program approach to the objectives indicated above. The sample language and forms are to be considered in the context of your own company and its manner of operation.

Carefully consider these objectives during the review and editing process of the sample section material in order to customize the section content to the specific needs of your organization. Prepare final language that is consistent with decisions made by your own management in the determination of your own company program.

It is important to understand that safety rules and regulations change from time to time. Be sure to review the final draft of the section and the entire program manual with appropriate safety, insurance, and legal professionals prior to issue and use. Periodically review your program with company management and all appropriate professionals to be certain that the company program remains current with all requirements.

COMPANY LOGO	Construction Safety and Loss Control Program Manual	Page: 13-1	Appr:
	13. NEW HIRE PERSONAL SAFETY PROGRAM	Revision Date:	

(Sample)

SECTION 13

(COMPANY) New-Hire Personal Safety Program

COMPANY LOGO	Construction Safety and Loss Control Program Manual	Page: 13-2	Appr:
	13. NEW HIRE PERSONAL SAFETY PROGRAM	Revision Date:	

13.1 THE (COMPANY) PERSONAL SAFETY PROGRAM

NOTE TO MANUAL PREPARER: The material that follows assumes the presence of a company Human Resources Department. If your company currently deals with human resource issues through any other method, that method must be applied to the sample section material.

The (**COMPANY**) Personal Safety Program of Section 13 is a condensed and summarized version of the more complete ***Safety and Loss Control Program Manual***.

It is to be provided to each new (**COMPANY**) employee upon hire as part of the complete New Hire package provided by the company Human Resources Department, and specifically reviewed in detail with the employee by the Human Resources Department before the new employee is released to begin any project assignment. It is provided in a separate booklet form for that application, and is repeated in this section in its entirely.

Each employee is to be made aware of the complete ***Safety and Loss Control Program Manual*** and its availability, and is to be encouraged to review the complete manual at every opportunity.

13.2 (COMPANY) SAFETY POLICY STATEMENT

As (**COMPANY**) continues through our (number) year of providing construction services, we've taken great pride in our consistent record of safety and our concern for the safety and health of all of our employees. There is no aspect of our work that takes on greater priority.

It is the policy of (**COMPANY**) to:

- Provide safe working environments.
- Comply with all laws.
- Never compromise a safe working condition for any reason.
- Conduct all operations in ways that reduce risk to all employees and to all other workers.
- Maintain conditions in ways that eliminate risk to visitors and the public.
- Eliminate risk of damage to property and equipment on and adjacent to each jobsite.

- Never accept any unsafe condition for any reason.

The results of the combined safety efforts of all employees will greatly affect the overall company success, as well as the manner in which the performance of all employees will be measured.

Our objective is 100% accident-free work, while maintaining the highest standards of quality construction. The cooperation and good intentions, and best judgment of all **(COMPANY)** employees in promoting safe and responsible work practices each day is the path toward continued personal and company improvement.

This ***Personal Safety Program*** is a condensed version of the complete ***Safety and Loss Control Program Manual*** that is in the possession of each Project Manager, Project Engineer, Site Superintendent, and Safety and Health Manager. The complete Manual defines the responsibilities and methods which will work to reduce and minimize personnel and economic losses, and to assist all employees to comply with all current laws and regulations. Please refer to the complete Manual at every opportunity.

13.3 RESPONSIBILITIES

13.3.1 Senior Management

NOTE TO MANUAL PREPARER: The sample subsections that follow are examples of summary descriptions of the responsibility of each employee as they apply to specific management decisions made and the way those decisions are accommodated in the respective sample section material. It is important to consider each sample subsection in this context, and edit the final draft so that it is accurate with respect to the actual decisions made by your management with respect to those issues.

Prevention of accidents is a complete management responsibility, and its highest priority. Senior management will support, encourage and enforce the Safety & Loss Control Program in order to ensure that it provides direction necessary to effectively implement all safety practices and procedures.

13.3.2 The Safety and Health Manager (Safety Officer)

The Safety and Health Manager is responsible for the implementation and general enforcement of the complete Safety & Loss Control Program. The basic responsibilities include:

COMPANY LOGO	Construction Safety and Loss Control Program Manual	Page: 13-4	Appr:
	13. NEW HIRE PERSONAL SAFETY PROGRAM	Revision Date:	

- Administer the Safety & Loss Control Program.
- Develop improvements in the Safety & Loss Control Program on an ongoing basis, and propose such improvements to senior management for approval.
- Establish and maintain procedures to identify unsafe acts and conditions before they cause accidents, and institute effective corrections.
- Provide adequately stocked first aid kits to all foremen, in all designated company vehicles, and at all jobsites. Develop a system of monitoring the supplies and keeping them all maintained.
- Participate in regular company training programs that will educate employees in accident prevention, safe work practices, hazard communication, and use of hazardous materials.
- Conduct a safety meeting at least quarterly with each crew.
- Conduct regular safety inspections at all jobsites and on all company vehicles in order to verify that the **(COMPANY)** safety program is being carried out properly.
- Maintain all required records of safety program information.

13.3.3 Project Managers and Site Superintendents

The Site Superintendents will have the primary responsibility for the safe conditions of their work areas. This includes protection of all employees directly under their direction, as well as the general public from injury, and protection of all company equipment and materials from damage.

Each Project Manager and Site Superintendent will be assigned his or her personal copy of the complete **(COMPANY)** *Safety and Loss Control Program Manual.* It will be the issued employee's responsibility to become thoroughly familiar with the Manual's contents, to maintain the Manual in a current and usable condition, and to conduct himself or herself strictly according to the guidelines presented therein.

13.3.4 Crew Foremen

All company foremen are to have direct daily contact with their crew members, regularly instructing each employee regarding the scope and safe manner of their specific work activities. The influence of the foreman is significant in the promotion of safety among all **(COMPANY)** employees under his or her direct control.

Responsibilities of the foreman include:

- Becoming thoroughly familiar with all requirements of the company Safety & Loss Control Program.
- Enforcing all safety rules and regulations within the scope of their work assignments.
- Initiating and implementing corrections to every unsafe act or unsafe condition in every work area under his or her control.
- Coordinating with the company Safety and Health Manager on matters relating to accident prevention and the safety of all (**COMPANY**) employees.
- Routinely monitoring their jobsites in an effort to detect unsafe acts or conditions that require correction.
- Monitoring each crew member's daily equipment maintenance and service procedures, in order to confirm safe operating condition of all equipment at all times.
- Establishing safe parking areas clear of equipment operation, prohibiting personnel from working unnecessarily close to equipment, and establishing job site traffic and pedestrian control measures.
- Explaining all applicable safe practice regulations and procedures to company employees under their direct supervision.
- Training employees under their direct supervision in safe and efficient ways to perform their assigned tasks.
- Regularly ensuring that all fire extinguishers are properly charged, and other appropriate safety equipment is maintained and stocked in fully operational condition.

13.3.5 Employees

All (**COMPANY**) employees not only have a responsibility to themselves for their own safety, but to their family, fellow employees, and employer as well.

All employees at a minimum shall:

- Be thoroughly aware of all company rules and policies including the company safety policy and comply with them in every respect.
- Comply with direction given to them by their supervisors.
- Store and maintain all equipment, tools, etc. in ways that are orderly, clean, and will ensure their continuing safe operation and use.

- Use all required personal protective equipment, such as hard hats, vests, and safety goggles wherever appropriate and required.
- Immediately report all accidents and injuries to their supervisor, no matter how slight in appearance.
- Report working conditions that are apparently unsafe, or suspected of being unsafe, to their supervisor.

13.4 JOBSITE AND SHOP WORK SAFETY RULES

NOTE TO MANUAL PREPARER: Although the sample copy below may appear to be reasonable in its apparent intent, it must be carefully considered with your own Human Resources Manager before finalizing the actual policy and language to be used in your actual Safety and Loss Control Program Manual.

13.4.1 General Description

The management of (**COMPANY**) will never intentionally allow a few individuals to jeopardize the rights and welfare of others by their willful disregard and careless behavior. Consequently, it is the policy of (**COMPANY**) that ***any of the following or similar types of violations shall lead to disciplinary action, and may lead to immediate termination of employment:***

1. Violations that will result in *immediate, automatic termination* include:
   - Consumption of alcoholic beverages during working hours.
   - Being under the influence of drugs or alcohol while on any jobsite or any (**COMPANY**) property.
   - Fighting and/or assaulting another employee, supervisor, client, or any other person.
   - Any willful destruction of company or other property or equipment.
   - Possession of a dangerous weapon while on any jobsite or company property, or any other location during working hours .
   - Careless, negligent, or reckless use of company equipment, machines, tools, or vehicles, or unauthorized use of (**COMPANY**) equipment.
2. Serious violations that will lead to severe disciplinary action and may lead to

2. Serious violations that will lead to severe disciplinary action and may lead to immediate or eventual termination of employment include:
   - Any type of horseplay.
   - Disregard of safety rules and/or other types of acts that endanger yourself or others.

13.4.2 General Procedures

1. Operators will ensure that safety shields and guards are in place and secured on all equipment prior to commencement of equipment operation.
2. Only approved shoring devices shall be used in all trenching operations as required by the Trench Safety Act. A "competent person" as defined by OSHA is required to be on site overseeing trenching operations.
3. Each foreman will be sure to have a completely stocked first aid kit at their jobsite(s).
4. The Safety and Health Manager is to maintain inventories of first aid supplies to be used to restock the jobsite first aid kits immediately upon the request of the foreman.
5. All employees shall wear protective hard hats at all times in all construction areas including any buildings, trenches, and open sites. The only exceptions are the jobsite field offices and any *finished* building areas.
6. Safety goggles must be worn while using power tools, sledge hammers, cutting tools, or while performing any task where there may be a danger of flying or falling particles which might cause eye injury.
7. Appropriate work shoes or boots must be worn on the job, preferably steel-toes and made of leather. Sneakers, tennis shoes, sandals, and similar footwear are *not* acceptable.
8. Fluorescent safety cones must be placed at and around open manholes or physical obstructions in order to prevent injuries or damage to vehicles.
9. Fluorescent orange vests must be worn by all field employees at all times when working on the site, outside of any building areas.
10. Gasoline shall be transported only in approved vapor release containers.
11. Materials and supplies identified under the HAZ-COM program shall be transported to the jobsite in the original packaging (i.e., chlorine, mortar and cement mixes, etc.), with the product date/safety labels intact.

13.4.3 Clothing

1. Only full-covered leather work shoes are to be worn. Sneakers, canvas shoes, or shoes that are open in any way are prohibited.

allowed.

3. Full-length pants must be worn at all times. Shorts are not allowed.
4. (**COMPANY**) will provide new rubber boots at a cost of (**$INSERT VALUE**) to the employee. The deduction will be made from the employee's first pay check. The boots will then belong to the employee, who will be responsible to keep then available for times when they are needed.
5. Any uniforms provided to employees will be at one-half the cost after the employee's probation period. Upon termination of employment with (**COMPANY**), employees are responsible for the return of all uniforms. The cost of uniforms not returned will be deducted from the employee's final paycheck.
6. Jewelry is not to be worn on any jobsite at any time. Rings, necklaces, and earrings may become caught on protruding objects or tangled in machinery. A watch may be worn unless the employee is performing any task which may result in the watch band being caught or in objects becoming lodged between the band and the skin. Watch bands should be of the expansion type so that the watch will slip off if it should get caught.
7. Welders and other operators of similar equipment are required to wear all full protection appropriate to the task being performed.

13.4.4 Vehicle Operation

The equipment, machines, and vehicles provided by (COMPANY) and its agents for use by any employee are to be used only for those tasks and purposes for which they were designed and issued. In addition:

- Company trucks must be checked by their drivers *each morning* before leaving the yard. All safety items such as headlights, tail and brake lights, horn, windshield wipers, and seat belts must be confirmed to be operative.
- Class "A" CDL licenses with hazard are required to operate all transport and fuel trucks. Class "B" may be acceptable under certain conditions, subject to the approval by the (**COMPANY**) Human Resources Department.
- Past driving records will be secured on all vehicle operators. Those with adverse histories will not be allowed to drive company vehicles.
- Only company trained or experienced operators are authorized to tow trailers.
- Drivers are responsible for checking their assigned vehicles out in the morning before leaving the shop. All maintenance items, including tires, oil, lights, horns, back-up alarms, etc. should be confirmed to be in working order, or must be

corrected immediately.

- All vehicles are equipped with fire extinguisher, first aid kit, and emergency equipment (flares, markers, etc.). The driver must confirm that all such items are present and adequately stocked.
- The driver must verify that any towed equipment should have two safety chains. These should be of adequate strength to maintain the towed object behind the towing vehicle in the event that it becomes uncoupled.
- The driver is responsible to verify that trucks are properly loaded at all times, and that loads are thoroughly and properly secured. Tarpaulins should be used to cover all loads on dump beds.
- The driver is responsible to verify that all flatbed trucks (dump, high-lifts, etc.) have tailgates or toeboards.
- The driver is responsible to verify that any L.P. (liquid propane) gas bottles are transported in an upright position, chained or fastened in a manner to prevent upset. Bottles must be capped, and all regulators, hoses, etc. are to be disconnected from the valves on the bottle.

13.4.5 Machinery and Mechanized Equipment

Preventive maintenance on all equipment is the responsibility of the equipment operator.

- It is the responsibility of the equipment operator to report any unsafe conditions of either the equipment or the jobsite to the foreman.
- Machinery and mechanized equipment shall be operated only by specifically designated personnel.
- Equipment without a designated operator, such as pumps, fall under the responsibility of the foreman for maintenance.
- Riding on equipment or loads is strictly prohibited, unless the equipment has been specifically designed for that purpose.
- ***Getting on and off moving equipment is prohibited under all circumstances, and may be grounds for immediate termination of employment.***

COMPANY LOGO	Construction Safety and Loss Control Program Manual	Page: 13-10	Appr:
	13. NEW HIRE PERSONAL SAFETY PROGRAM	Revision Date:	

13.5 REPORTING PROCEDURES

All company employees are responsible for the following:

- Immediately report all unsafe tools and equipment to your foreman or superintendent.
- Immediately report all injuries and accidents to your immediate supervisor—however slight the accident or injury may appear to be.
- Immediately report all hazardous conditions to your foreman or superintendent.

13.6 DISCIPLINARY ACTION

Any employee found to be in any violation of a company safety rule will at a minimum be given a written warning by his or her supervisor. In the case of a serious violation as determined by **(COMPANY)** supervision, the employee may be sent home for the day without pay.

A second violation of the same rule, or extremely serious violation of any rule, will be grounds for immediate termination of employment.

13.7 CONCLUSION

The safety of the employee, the safety of all fellow employees, the safety of any other worker or supervisor on the jobsite, and the safety of the general public is the utmost concern of **(COMPANY)** and its management. We expect the full cooperation of every employee, and the active participation by each employee in making every **(COMPANY)** jobsite, shop area, and office the safest environment that it can be.

Sample Section 14

ADMINISTRATIVE LOGS, FORMS, AND RECORDS

Section Use

The form and content of Sample Section 14 *Administrative Logs, Forms, and Records* is designed as an example of company instruction to its employees with respect to the material presented. It is intended to provide for each employee:

- Each form available to be used for the self-explanatory purposes intended to accommodate various administrative functions, legal compliances, and business necessities.
- A consolidated location of each form, to facilitate immediate availability for their use when needed.

The sample section that follows is an example of a company safety and loss control program approach to the objectives indicated above. The sample language and forms are to be considered in the context of your own company and its manner of operation.

The manual preparer must carefully consider the complete format, content, and use of each form as described specifically in the context of your own company operation. Add, delete, and modify all forms as appropriate, in order to keep your own Safety and Loss Control Program Manual directly relevant to your own company.

It is important to understand that safety rules and regulations change from time to time. Be sure to review the final draft of the section and the entire program manual with appropriate safety, insurance, and legal professionals prior to issue and use. Periodically review your program with company management and all appropriate professionals to be certain that the company program remains current with all requirements.

COMPANY LOGO	Construction Safety and Loss Control Program Manual	Page: 14-1	Appr:
	14. ADMINISTRATIVE LOGS, FORMS, AND RECORDS	Revision Date:	

(Sample)

SECTION 14

Administrative Logs, Forms, and Records

COMPANY LOGO	Construction Safety and Loss Control Program Manual	Page: 14-2	Appr:
	14. ADMINISTRATIVE LOGS, FORMS, AND RECORDS	Revision Date:	

14.1 DEVELOPMENT AND USE OF FORMS

NOTE TO MANUAL PREPARER: Consider the sample forms in this section and their intended use in the context of your own specific operation. Modify each form as appropriate to accommodate your own documentation and information distribution systems as well.

Note additionally that each form addresses not only the safety and loss control objectives of the organization, but necessarily addresses multiple legal considerations. As such, it is critically important to have the specific language, content, and manner of presentation of each form reviewed by competent legal, insurance, and safety professionals prior to implementation.

The forms that follow have been developed for use by (**COMPANY**) employees in establishing and implementing the procedures developed in the (**COMPANY**) Safety & Loss Control Program.

Use the forms according to the descriptions and instruction below. Contact the (**COMPANY**) Safety and Health Manager if there are any questions.

14.2 SAFETY INSPECTION RECORD

The *Safety Inspection Record* is used to facilitate any routine or special jobsite safety review or inspection by the (**COMPANY**) Safety and Health Manager, Project Manager, Site Superintendent, or any other (**COMPANY**) employee authorized to perform such inspections.

Use the ***Jobsite Safety Review Checklist*** of Section 4.8 of the ***Safety and Loss Control Program Manual*** as an aid in identifying actual or potential problems, and use this ***Safety Inspection Record*** to record relevant observations.

The record can be used to document unsafe conditions by (**COMPANY**) or other subcontractors, trade contractors, the general contractor, construction manager, and by (**COMPANY**)'s own forces. It is vital that timely distribution of the record be made as indicated on the bottom of the form. A reply should be made by the observed offending party(ies) to any remarks or recommendations given.

COMPANY LOGO	Construction Safety and Loss Control Program Manual	Page: 14-3	Appr:
	14. ADMINISTRATIVE LOGS, FORMS, AND RECORDS	Revision Date:	

14.2 Sample Safety Inspection Record

EMPLOYER: ______________________________

Inspection Date: ____________ Time: ____________ ☐ AM ☐ PM

Project:
Name: ____________________ No.: ____________
Location: ______________________________

USE ADDITIONAL FORMS AS MAY BE REQUIRED FOR A COMPLETE INSPECTION

1. Unsafe conditions observed on the Jobsite:

Unsafe conditions observed on the Jobsite:	Individual/company notified of the violation:	Date Notified:
		/ /
		/ /
		/ /
		/ /
		/ /
		/ /
		/ /
		/ /
		/ /
		/ /
		/ /
		/ /
		/ /
		/ /
		/ /
		/ /
		/ /

2. Unsafe acts performed by individual(s) on the jobsite

Unsafe acts performed by individual(s) on the jobsite	Individual/company notified of the violation:	Date Notified:
		/ /
		/ /
		/ /
		/ /
		/ /
		/ /
		/ /
		/ /
		/ /
		/ /
		/ /
		/ /
		/ /
		/ /
		/ /
		/ /

4. Inspection Conducted by: ______________________ Title: ______________

cc: ☐ Safety & Health Manager ☐ Project Manager ☐ ____________
☐ Human Resources Manager ☐ ____________ ☐ ____________
☐ Every Individual & Company Named Above.

14.3 SAFETY TRAINING FORM

Project Managers, Superintendents, and Foremen can use this *Safety Training Log* to document Tailgate Safety Meetings, or other training being done on a (**COMPANY**) project. Generally, it is intended for use by (**COMPANY**), but it may also be used for subcontractors if situations permit. Any use for other than (**COMPANY**) employees must be approved by the Safety and Health Manager.

COMPANY LOGO	Construction Safety and Loss Control Program Manual	Page: 14-5	Appr:
	14. ADMINISTRATIVE LOGS, FORMS, AND RECORDS	Revision Date:	

14.3 Sample Safety Training Log

EMPLOYER: ____________________

Training Date: ____________________ Time: ____________ ☐ AM ☐ PM

Project:
Name: ____________________ No.: ________
Location: ____________________

Training Conducted by: *(Name)* ____________________ ☐ Company Employee: *(Title)* ____________________
☐ Other: ____________________

Training Aids Used: ☐ Handout materials given: ____________________
☐ Audio/visual Materials Used: ____________________

Training Meeting Attendees *(Circulate this form for signatures):*

	Name *(Print)*	Signature	Company Employee
1.			☐ YES ☐ NO
2.			☐ YES ☐ NO
3.			☐ YES ☐ NO
4.			☐ YES ☐ NO
5.			☐ YES ☐ NO
6.			☐ YES ☐ NO
7			☐ YES ☐ NO
8.			☐ YES ☐ NO
9.			☐ YES ☐ NO
10.			☐ YES ☐ NO
11.			☐ YES ☐ NO
12.			☐ YES ☐ NO
13.			☐ YES ☐ NO
14.			☐ YES ☐ NO
15.			☐ YES ☐ NO
16.			☐ YES ☐ NO
17.			☐ YES ☐ NO
18.			☐ YES ☐ NO
19.			☐ YES ☐ NO
20.			☐ YES ☐ NO

Description of Training / Remarks:

Safety Meeting Minutes Attached: ☐ YES ☐ NO

cc: ☐ Safety & Health Manager ☐ Project Manager ☐ ____________
☐ Human Resources Manager ☐ ____________ ☐ ____________

14.4 SAFETY MEETING MINUTES FORM

This *Safety Meeting Minutes Form* is to be used to facilitate recording, filing, and record distribution of the content of each safety meeting. If the project needs require a separate memo or report regarding safety communication, this form may also be used to meet those needs.

The format should also include all the headings on the Safety Meeting Minutes form. Note that as record of attendance, a sign-in sheet that includes respective company representatives should be attached to the minutes of every Safety Meeting at a minimum. Preferably, however, the Safety Training Log of Section 14.3 should be used for this purpose.

14.5 SAFETY VIOLATION NOTICE

NOTE TO MANUAL PREPARER: The language of the sample section that follows correlates with the sample notice form included in the following pages. If the form itself is modified or not used for any reason, the language below must be changed accordingly.

The *Safety Violation Notice* is to be used by (**COMPANY**) supervision to identify observed safety violations, to record the observation, and to indicate appropriate correction. It is to be delivered to subcontractors who refuse to correct a noted safety violation, or who do not correct a noted safety violation within a reasonable period of time, and is to be distributed to appropriate individuals and files for important record purposes.

The *Safety Violation Notice* is vital documentation that will support the company's position whenever responsibility for a safety violation will need to be confirmed. It will serve as written notice that:

- Such violations must be corrected,
- The date by which they must be corrected,
- Failure to correct will result in corrective measures being taken by (**COMPANY**), and
- The Subcontractor will be held responsible for all resulting costs, delays, damages, and other effects.

Each Notice should be kept in a current file and periodically reviewed. Current and outstanding notices should be an important agenda item on upcoming Safety Meetings of every type.

When a violation is corrected, it should be so noted on the Notice, and maintained in the offending party's file.

COMPANY LOGO	Construction Safety and Loss Control Program Manual	Page: 14-8	Appr:
	14. ADMINISTRATIVE LOGS, FORMS, AND RECORDS	Revision Date:	

14.4 Sample Safety Meeting Minutes Form

COMPANY: ______________________________

Meeting Date: ____________________ Time: ____________________ ☐ AM ☐ PM

Project:
Name: ______________________________ No.: ____________
Location: ______________________________

USE ADDITIONAL FORMS OR PAGES TO COMPLETE ANY SECTION BELOW AS MAY BE REQUIRED

1. Safety Training Log dated _____/_____/_____ is attached.

2. Summary of Tailgate Safety Meetings conducted since the last periodic Safety Meeting:

   Date: _____/_____/_____ Subject: ______________________________

   Date: _____/_____/_____ Subject: ______________________________

   Date: _____/_____/_____ Subject: ______________________________

   Date: _____/_____/_____ Subject: ______________________________

   Date: _____/_____/_____ Subject: ______________________________

3. Accidents that have occurred since the last periodic Safety Meeting:

4. Measures that have been taken to avoid recurrence of these types of accidents:

5. Current possible hazards observed on the jobsite:

6. Remarks:

7. Meeting Conducted by: ______________________________ Title: ______________________________

cc: ☐ Safety & Health Manager ☐ Project Manager ☐ ____________

☐ Human Resources Manager ☐ ____________ ☐ ____________

COMPANY LOGO	Construction Safety and Loss Control Program Manual	Page: 14-9	Appr:
	14. ADMINISTRATIVE LOGS, FORMS, AND RECORDS	Revision Date:	

14.5 Sample Safety Violation Notice

(COMPANY) Safety Violation Notice

Date of Notice: _____/_____/_____ Time: _____:_____ AM PM

Date of Violation: _____/_____/_____ Time: _____:_____ AM PM

1. Individual:
   Name: ____________________ Title: ____________________

2. ☐ (COMPANY) Employee
   ☐ Subcontractor/Suppplier: ____________________

3. Description of Violation:
   ____
   ____
   ____
   ____
   ____
   ____
   ____
   ____
   ____
   ____

4. Corrective Action Required ☐ IMMEDIATELY ☐ By _____/_____/_____ :
   ____
   ____
   ____
   ____
   ____
   ____
   ____
   ____
   ____
   ____

5. Company Employee warned of: ☐ Suspension ☐ Dismissal ☐ Dismissal Without Pay ☐ First Offense / No Warning

6. Notice Prepared by: ____________________ Title: ____________________

cc: ☐ Safety & Health Manager ☐ Project Manager ☐ ____________________
☐ Human Resources Manager ☐ ____________________ ☐ ____________________

14.6 HOT WORK PERMIT

NOTE TO MANUAL PREPARER: The sample policy described below is considered very differently in different companies and in different geographic locations. Consider the self-explanatory statements as well as the specific language on the sample form itself and make your own decision with respect to the manner in which you intend your own company to deal with the respective conditions.

The *Hot Work Permit Form* is to be used whenever any type of friendly fire is intended to be used on any jobsite under (**COMPANY**) control. This includes welding, cutting, burning, and similar operations that may result in open flame, sparks, hot slag, production of combustible gasses, and similar conditions.

A Separate Permit is required for each hot work operation each day.

Approximately two hours after completion of the hot work, the area is to be inspected to assure the absence of fire.

COMPANY LOGO	Construction Safety and Loss Control Program Manual	Page: 14-11	Appr:
	14. ADMINISTRATIVE LOGS, FORMS, AND RECORDS	Revision Date:	

14.6 Sample Hot Work Permit Form

(COMPANY) Hot Work Permit

Date of Permit: _____/_____/_____ Time: _____:_____ AM PM

Permit Duration: **FROM:**
_____/_____/_____ Time: _____:_____ AM PM

TO:
_____/_____/_____ Time: _____:_____ AM PM

Project:
Name: ____ No.:_______________
Location: ____

ALL COMBUSTIBLE MATERIALS ARE TO BE REMOVED FROM THE WORK AREA AND ITS PERIMETER

1. Description of Work Performed:

☐ WELDING ☐ TORCH CUTTING ☐ SOLDERING ☐ OTHER

Describe: ____

2. Work Area(s) Affected:

3. **FIRE WATCH EQUIPPED WITH:**

PERMIT EXPIRES AT END OF WORK DAY ON DATE INDICATED ABOVE

Prepared by: ______________________________ _____ Title: ________________________

cc: ☐ Safety & Health Manager ☐ Project Manager ☐ ____________________
☐ Human Resources Manager ☐ ____________________ ☐ ____________________

14.7 JOBSITE VISITOR SIGN-IN AND GENERAL RELEASE

All visitors to any (**COMPANY**) jobsite must be required to sign the *Jobsite Visitor Sign-In and General Release* before being allowed to proceed anywhere on the jobsite. There should be a ready supply of these forms maintained at the jobsite field office visitor sign-in location. It is often helpful to maintain the supply on a clip board hung adjacent to the visitor sign-in clip board.

The heading (including all project information) should be completed in advance by the Project Manager, so that all a visitor needs to do is add their name at the bottom.

If any visitor to the jobsite refuses to sign the form, the visitor is to be advised that he or she is not allowed beyond the (**COMPANY**) field office. If the visitor proceeds into the jobsite in violation of this restriction, contact the (**COMPANY**) Safety and Health Manager immediately for appropriate instruction.

14.7 Sample Jobsite Visitor Sign-In and General Release

COMPANY: ______________________________

Project:

Name: ______________________________ No.: __________

Location: ______________________________

THE UNDERSIGNED hereby acknowledges and agrees that permission to enter upon the premises of the Project indicated above for any purposes whatsoever, including observation, inspection, delivery or materials and/or equipment, posting notices, delivering information of any type, performing any work, or for any other reason whatsoever is granted by the company named above as a courtesy by sole and exclusive benefit. I hereby attest that in consideration of such permission I agree for my self any my heirs that I release and forever discharge the company named above from any and all claims, actions, or causes of action of any kind which I or my heirs now have or may have at any time hereafter related in any way to death, bodily injury, or damage to property which may occur to me and/or my property while on the premises of the Project, whether or not such death, injury, or property damage is or may be due wholly or in part to any act or negligence on the part of the above named company. Accordingly, I fully understand, agree with, and accept full responsibility for all risks associated with the above.

Name (Print):	Signature:	Date:

Sample Section 15

SAFETY AND LOSS CONTROL ON THE JOBSITE

Section Use

The form and content of Sample Section 15 *Safety and Loss Control on the Jobsite* is designed as an example of company instruction to its employees with respect to the material presented. It is intended to provide for each employee specific instruction in dealing with particular issues and situations that may arise on a jobsite. Note that:

- The itemized instruction is arranged alphabetically, in order to facilitate quick research and clear direction, and
- If desired, the entire sample section can be reproduced as a separate field manual and coordinated with the company's actual Safety & Loss Control Program Manual. In this way, it may be possible to highlight the importance of the need to refer to the information regularly and as needed.

The sample section that follows is an example of a company safety and loss control program approach to the objectives indicated above. The sample language and forms are to be considered in the context of your own company and its manner of operation.

Carefully consider these objectives during the review and editing process of the section material in order to customize the section content to the specific needs of your organization. Prepare final language that is consistent with decisions made by your own management in the determination of your own company program.

It is important to recognize the possibility of changes in OSHA regulations, policies, or procedures. It is critical that the manual preparer confirm all remarks of this section and all other sections of this *Safety and Loss Control Program Manual* made with regard to such information and direction that have been presented as an example of how to communicate this kind of information to company employees.

It is also important to understand that safety rules and regulations change from time to time. Be sure to review the final draft of the section and the entire program manual with appropriate safety, insurance, and legal professionals prior to issue and use. Periodically review your program with company management and all appropriate professionals to be certain that the company program remains current with all requirements.

COMPANY LOGO	Construction Safety and Loss Control Program Manual	Page: 15-1	Appr:
	15. SAFETY AND LOSS CONTROL ON THE JOBSITE	Revision Date:	

(Sample)

SECTION 15

Safety and Loss Control on the Jobsite

COMPANY LOGO	Construction Safety and Loss Control Program Manual	Page: 15-3	Appr:
	15. SAFETY AND LOSS CONTROL ON THE JOBSITE	Revision Date:	

COMPANY LOGO	Construction Safety and Loss Control Program Manual	Page: 15-4	Appr:
	15. SAFETY AND LOSS CONTROL ON THE JOBSITE	Revision Date:	

15.1 SECTION DESCRIPTION

NOTE TO MANUAL PREPARER: This entire Sample Section 15 can be left as it is within the complete Safety and Loss Control Program Manual,* or *can be arranged as a separate submanual to be made available to all company field employees in order to improve the probability that the information will be referred to as it is needed, or can be provided in both places.

However the final arrangement is determined, be certain to explain it thoroughly in the following "Section Description" portion of the complete Safety and Loss Control Program Manual.

This section addresses particular field conditions, and describes the procedures to be followed by every (**COMPANY**) employee on every jobsite. It is arranged alphabetically for ease of reference.

The section information catalogs specific jobsite and shop hazards, and provides direct instruction that will help each employee to work in and around each condition in a safe manner that minimizes risk of injury or health impairment.

Each set of instructions has been written with the explicit intention of providing for the proper implementation of the respective OSHA and other legal and practical requirements with respect to the particular issue. In many cases, interpretation of the controlling standard or standards may have been necessary in order to determine an appropriate set of procedures to apply to or comply with the stated or implicit interpretation of the respective requirement.

Although every effort has been made to be comprehensive, complete, and practical, it is possible that a condition on a jobsite may not be completely addressed, or a new condition may be encountered which may not be covered at all. Each new situation of these types must be immediately addressed by referring to and applying local, state, and/or federal laws governing the process. In such cases, the (**COMPANY**) Safety and Health Manager should be immediately contacted for assistance and direction.

The company Safety and Health Manager has the authority to halt any work activity which he or she believes may hold an imminent threat to life, health, property, or equipment.

NOTE TO MANUAL PREPARER: The balance of this section applies both interpretation of OSHA and other controlling legal standards, and examples of policies adopted by this sample Safety and Loss Control Program Manual. Consider each subsection carefully with respect to these ideas, and modify each subsection as you may determine to be appropriate to accommodate your specific situation, and your own company intentions.

15.2 BLASTING AND EXPLOSIVES

15.2.1 General Policy

a. No **(COMPANY)** employee is permitted to handle or use explosives or blasting materials.
b. All blasting and explosive work will be done by the use of outside contractors licensed and qualified to perform in the state in which the blasting and explosive work is to be performed.
c. All outside contractors and subcontractors engaged to perform blasting and explosive work must be required to provide all appropriate bonds and insurances as are necessary for proper indemnification of **(COMPANY)** throughout the performance of all blasting and explosive work. Confirm all such requirements with the **(COMPANY)** contract manager if necessary.
d. Smoking or open flames are not permitted within 50 ft. of explosives, detonator storage magazines, or other such materials and equipment.
e. When planning to perform any blasting near or around buildings or other structures, contact the Project Manager to determine the reasonable level of monitoring to confirm the existing and subsequent status of structures in the proximity of the blasting area. Such monitoring is critical to substantiating facts relating to subsequent claims by property owners for cracks or other damage alleged to have occurred as a result of the blasting activity.

15.3 COMPRESSED GASSES

15.3.1 Compressed Gas Cylinders

a. All cylinder valves must be closed when any work is finished and when any cylinders are empty or being moved.
b. Valve protection caps must be placed and secured properly before compressed gas cylinders are transported, moved, or stored.
c. Compressed gas cylinders must be secured in an upright position with chain or other appropriate and approved systems designed for such purpose at all times.
d. All cylinders must be kept a safe distance from welding or cutting operations or shielded from same.

COMPANY LOGO	Construction Safety and Loss Control Program Manual	Page: 15-7	Appr:
	15. SAFETY AND LOSS CONTROL ON THE JOBSITE	Revision Date:	

e. All cylinders must be placed where they cannot become part of an electrical circuit.
f. ***Oxygen and acetylene must not be stored together.***
g. Oxygen and fuel gas regulators, hoses, and associated equipment must never be modified from their factory design, and must be in proper working order while in use.

15.3.2 Compressed Air Cleaning

a. Compressed air can be extremely dangerous if allowed to penetrate the skin. Careless use of compressed air may also cause unnecessary dust and airborne particulates to be introduced into the work area atmospheres, thereby increasing risk of health danger to employees.
b. Compressed air used for cleaning purposes may not exceed 30 psi and then only in conjunction with effective chip guarding and personnel protective equipment that is designed for that application. Exceptions to 30 psi are allowed only for concrete form, mill scale, and similar cleaning operations.
c. ***The use of compressed air to clean off yourself or other workers is strictly prohibited.***

15.4 CONCRETE, CONCRETE FORMWORK, AND SHORING

15.4.1 Reinforcing Steel

a. All concrete reinforcing steel that leaves open ends or otherwise protrudes in a way that leaves a condition where employees could fall and be injured by cutting or impalement on the reinforcing steel bars must have protective devices put in place on the reinforcing steel, or otherwise be guarded to eliminate the hazard of cutting or impalement.
b. Welded Wire Mesh (or Welded Wire Fabric) is to be placed in a manner that prevents or otherwise is secured from recoiling.

15.4.2 Formwork and Shoring

a. Concrete formwork and shoring is to be designed and constructed in a manner that safely supports all loads imposed during concrete placement. All formwork and shoring is to be designed strictly in accordance with published

COMPANY LOGO	Construction Safety and Loss Control Program Manual	Page: 15-8	Appr:
	15. SAFETY AND LOSS CONTROL ON THE JOBSITE	Revision Date:	

practices of the formwork manufacturer, and performed by qualified personnel. All formwork and shoring components are to be inspected by qualified personnel prior to, during, and after erection. Drawings or plans of jack layout, formwork, shoring, working decks and scaffolding systems are to be available at the jobsite at all times.

b. Formwork and shoring are not to be removed until it has been determined through proper testing that the concrete has gained sufficient strength to support its weight and any superimposed loads, and meets all requirements of the project contract documents.

15.5 CRANES AND HOISTING EQUIPMENT

15.5.1 General Procedures

a. On every crane or piece of hoisting equipment notices of all rated load capacities, recommended operating speeds, and any hazard warnings or special instructions must be conspicuously posted. All instructions and warnings must be visible from the equipment operator's station.
b. Copies of maintenance and inspection records must be presented in the operator's cab if there is a practical area for appropriate storage. If not, contact the **(COMPANY)** Equipment Manager for instruction regarding the maintenance and storage of such records.
c. All areas within swing radius of cranes that are potentially accessible by pedestrian, vehicular, or equipment movement must be barricaded to prevent anyone or any vehicle or equipment from being struck by the crane or hoisting equipment, or its load(s).
d. Except where electrical distribution and transmission lines have been de-energized and visibly grounded, or where insulating barriers not a part of or an attachment to the equipment or machinery have been erected to prevent physical contact with such lines, no part of a crane or its load shall be operated within 15 ft. of any power line.
e. Cranes will be inspected before each use by the operator. Any defects must be corrected before use. Logs of crane inspection must be kept with the crane.

15.5.2 Suspended Personnel Platforms

a. Although use of suspended personnel platforms is allowed as specified in OSHA 1926.550, it is the policy of **(COMPANY)** that suspended personnel platforms with crane or hoisting equipment not be used unless the provision and use of conventional means of reaching the worksite would be more hazardous or not possible. It is recognized, however, that such a circumstance is highly unusual, and will only be considered by the company in extreme and unique situations.

15.6 DISPOSAL CHUTES

a. An enclosed chute is to be used whenever materials are dropped more than 20 ft. to any exterior point of a building or site area.
b. When debris is dropped through floor holes without a chute, the area where the material is dropped must be enclosed with barricades at a minimum of 42 in. and not less than 6 ft. back from projected edges of the opening overhead.
c. Warning signs are to be posted at each floor level.

15.7 ELECTRICAL

15.7.1 Permanent Electrical Installations

a. All permanent installations of electrical materials, equipment, and circuitry or repairs, modifications, or maintenance to permanent installations of electrical materials, equipment, or circuitry must be done by a qualified person properly licensed as required in the location in which the work is performed. The individuals performing the work must meet all applicable federal, state, and local codes and requirements.
b. All repairs, modifications, and maintenance required on permanent installations of electrical materials, equipment, and circuitry must meet or exceed their original capacity and quality.

15.7.2 Temporary Electrical Installations and Uses During Construction Activity

The following guidelines are to be considered the absolute minimum requirements to be supplemented by requirements of all applicable codes and standards for such work:

a. Wiring and Branch Circuits
   i. Must be protected by a proper amperage overcurrent device such as a fuse or circuit breaker.

ii. Must be located so as to prevent physical damage to the wire conductors.

b. Lighting
   i. Temporary lights must be located at appropriate distances from the respective lighted work area(s).
   ii. All temporary lighting fixtures must be protected from accidental contact or breakage.
   iii. Temporary lights shall not be suspended by their electric cords unless they are designed for this type of installation.
   iv. Portable electric lights used in wet or potentially wet locations must be either low voltage type (12 volts or less) or protected by a GFI (ground fault interrupter).
   v. Portable light cords designed with (2) conductors are not to be used as extension cords to supply power tools.

c. Extension Cords
   i. Must be visually checked before each use and periodically while in use to assure their original integrity is maintained. Cords with cuts, breaks, deep abrasions, etc. shall be taken out of service immediately. Repairs to extension chords shall only be performed by qualified, licensed electricians.
   ii. Are not to be hung across walkways, work areas, stairways, etc. where they create possible tripping hazards, or where they might be damaged by equipment or traffic. Cords may be hung overhead with sufficient minimum clearance or rerouted under or around walkways and work places. Cords crossing vehicle traffic areas may be covered or encased by suitable material to protect them that is designed for this purpose.
   iii. Must not be fastened, or attached with any type of conductive material such as nails, metal wire, staples, etc.
   iv. Must not be allowed to lay in wet or potentially wet areas.
   v. Must be of the 3–wire type when used with portable electric tools and appliances.

15.7.3 Underground Electric

a. Before any work is allowed to begin in areas where there is a possibility that underground electric lines may be encountered, the senior company employee on the site must ensure that the necessary authorities are contacted, and the lines are located and marked.

b. Every electric line or cable of unknown origin that is discovered or exposed during a digging, drilling, probing, or similar operation is to be considered as energized and life threatening. The senior company employee on the site will ensure that all necessary safety precautions are taken in order to isolate the line from all workers and the public. Such precautions may include halting the operation if appropriate. The senior company employee on the site is to then contact the proper authorities to have the line identified and either confirmed to be abandoned and/or made safe for continuing the work.

c. In work areas where exact locations of suspected underground lines cannot be confirmed, employees using jack-hammers, probes, shovels, and similar hand tools must use tools made to insulate the user against electrical shock or must use insulated protective gloves designed for such use.

d. Any and all underground lines that are discovered or become severed must be considered energized on both sides of the sever, and be treated accordingly.

15.7.4 Overhead Electric

a. All overhead electric lines are to be considered as energized and life threatening until or unless it is positively confirmed that the proper authorities have identified and have deenergized them.

b. All overhead ground lines that are discovered or become severed must be considered energized on both sides of the sever, and be treated accordingly.

c. In general, equipment or machinery being moved or transported must maintain minimum clearances of 25 ft. to all power lines. If equipment or machinery must operate within 25 ft. of energized power lines, *one* of the following must occur before beginning work:

   i. The senior company employee on the site must positively confirm the voltage(s) of the line(s) involved and coordinate the following information with the (**COMPANY**) Safety and Health Manager to confirm the minimum clearances for the lines:

Line Voltages	**Minimum Clearance**
50 KV and less	10 ft.
51 KV to 200 KV	15 ft.
201 KV to 345 KV	20 ft.
346 KV to 500 KV	25 ft.

NOTE:

a) In lieu of the above clearances, a distance of at least twice the length of the line insulator may be used, but never less than 10 ft.

b) Equipment swing and/or boom restraints or alarms may be installed to physically limit or warn against proper clearance encroachment.

ii. If the operation of the project allows, the senior company employee on the site may contact the proper authorities and arrange to have the line(s) be deenergized, including physical grounds, for the duration of the operation.

iii. Arrangements are made with the proper authorities to have all such power line(s) effectively shielded during the construction operation.

c) The senior company employee on the site is to designate a person to assist the equipment operator in observing and maintaining proper clearances in order to comply with the above.

d) An adequate ground line or jumper shall be placed as needed in order to dissipate any induced voltage charge (static electricity) in equipment or materials being handled in the vicinity of transmitter towers or high voltage electric lines. In addition, combustible and flammable liquids are not allowed to be stored or transported through such areas.

15.7.5 Electrical Grounding Program for 110 Volt AC Equipment

NOTE TO MANUAL PREPARER: The procedure described below is self explanatory, but anticipates a specific procedure. Review the procedure described in order to determine if it is consistent with the procedure intended by your company. Modify the language as appropriate for the development of your own Safety and Loss Control Program Manual.

a. The electrical grounding program is established to provide a systematic inspection procedure for all 110 volt electrical equipment used on every company jobsite and shop area. The purpose of this inspection procedure is to identify any existing faults or potential faults with the grounding conductors of all 110 volt construction equipment, identify work area conditions that may cause or contribute to a fault condition, and implement immediate corrections to each situation identified. These efforts are necessary in order to prevent accidental electrical shock to an employee using the 110 volt electrical equipment. This program does not in any way replace regular safe-operating practice or the responsibility of all employees to visually check this and any other type of equipment before using it or placing it into operation.

b. Types of 110 volt equipment to be inspected in the program include: power tools, extension cords, light cords that have receptacles, plug and wiring adapters, multi-receptacle adapters, and portable generators with receptacles having three (3) wires. Exceptions are: double-insulated power tools, non-portable 110 volt equipment, receptacles that are part of a building or permanent structure, non-construction use equipment and equipment other than nominal

110 volt. Any 110 volt equipment that is always operated from a GFI (Ground Fault Interrupter) circuit is also exempt from the following Ground Assurance Program.

c. Electrical grounding inspections will be conducted periodically for equipment on all jobsites and shop areas, and will be performed by qualified personnel who have been trained in proper procedure. A grounding inspection schedule will be prepared by the company Safety and Health Manager and sent to the appropriate foreman and superintendent prior to each inspection. As much notice as possible will be given, advising each crew of the date they are to bring their equipment into the shop. Each crew is to bring their equipment into the shop at the end of their shift and pick it up the following work day prior to the beginning of their shift. Any equipment failing the inspection and not repaired or replaced will be noted to the foreman and other arrangements to secure replacement equipment will be made.

d. Any equipment required to be inspected but not available or feasible to be brought in at the scheduled time will be rescheduled by the shop for later inspection or on-site inspection as necessary.

e. Equipment under this program will be inspected quarterly. Color codes of tape will be used to identify equipment that has been inspected and passed for the respective quarters. The colored tape will be placed by the inspector around the cord near the plug end of each piece of equipment passing each periodic inspection according to the following scheme:

January–March	White (Winter)
April–June	Green (Spring)
July–September	Red (Summer)
October–December	Orange (Fall)

15.7.6 Battery Storing and Charging

a. Batteries of the unsealed type are to be stored in well-ventilated rooms or in enclosures with outside vents.

b. Batteries being charged must be in a well-ventilated open or outside area. Vent caps must be functioning properly and in place. Fumes must not be allowed to migrate into confined rooms or spaces.

c. There must always be a means in the immediate vicinity of charging operations to adequately wash out eyes and flush skin with water. An appropriate fire extinguisher must also be located in the vicinity of all battery charging operations.

15.8 FIRE PROTECTION, PREVENTION, AND EMERGENCY PROCEDURES

15.8.1 Fire Protection

NOTE TO MANUAL PREPARER: Confirm all respective current legal requirements before finalizing the language of the following sample section.

a. The basic first defense against fire is the portable fire extinguisher. Most commonly, ABC (universal extinguishing agent) type extinguishers in 2 to 20 lb. sizes are used in jobsite and shop areas, along with a smaller number of 15 to 20 lb. CO_2 BC (liquid or electrical class fire) type extinguishers for special circumstances.
b. Fire-fighting equipment must be conspicuously located and readily accessible at all times, and periodically inspected and maintained in operating condition. Immediately report any inoperative or missing equipment to jobsite and/or company supervision.
c. If a building project under construction includes automatic sprinkler protection, installation of the fire protection system will closely follow construction and be placed in service as soon as proper construction sequencing and applicable laws permit, following completion of each construction area.
d. On building jobsites, fire extinguishers, rated not less than 2A, will be provided for each 3,000 square ft. of building area (or major fraction). Travel distance from any point to the nearest fire extinguisher may not exceed 100 ft. with at least one extinguisher per floor. In multi-story buildings, at least one fire extinguisher must be located adjacent to each stairway.
e. Periodic fire protection/prevention training will be conducted under the direction of the company Safety and Health Manager, including instruction and demonstrations on the proper use of portable extinguishers. It is, however, the responsibility of each employee to read the instructions and familiarize themselves with the extinguisher(s) in their respective areas of work, whether it be in an office, shop area, construction area, on a piece of heavy equipment, or in a pick-up truck, or other company vehicle.
f. The extinguishers on each jobsite will be procured by the jobsite management and staff, and be maintained in the field by company jobsite personnel. Fire extinguisher quantities and areas of placement are subject to the direction of

the company Safety and Health Manager. Periodic and annual maintenance and inspection of all extinguishers will be controlled by the shop, either through an outside certified contractor or under the direction of the company Safety and Health Manager.

g. Should any company employee use an extinguisher, or observe one that has been used or that needs charge or other attention, they should immediately contact the shop or Safety and Health Manager. At a minimum, extinguishers requiring attention include those with gauges indicating low charge, gauges missing pins, missing inspection labels or tags, and obvious signs of damage or deterioration.
h. Access to all fire-fighting equipment must be kept clear at all times.
i. All heavy equipment and certain selected company vehicles will be equipped with a fire extinguisher.

15.8.2 Fire Prevention

a) General:
   i. Housekeeping on all company jobsites, in all offices, in all shop areas, in all field offices, and in all work trailers will be maintained in order to assure that no potential fire hazards exist.
   ii. Used cleaning rags that are oily, greasy, and/or fuel soaked are to be placed and stored in non-combustible solid containers with lids.
b) Outdoor Storage Areas
   i. Combustible materials are not to be piled higher than 20 ft. and shall have accessways of at least 15 ft. in width between piles and are to be placed at a minimum of 10 ft. from any building or structure.
   ii. Proper fire extinguishers shall be placed in combustible material storage yards so that no more than 100 ft. exist from an employee to the nearest extinguisher at any time.
c) Indoor Storage Areas
   i. The manner of material storage must not be allowed to obstruct or adversely affect any fire-fighting equipment or any means of exit.
   ii. Adequate aisle space must be maintained between all stored materials in order to accommodate the largest piece of fire-fighting equipment anticipated to be used in an area.
   iii. No material may be stored within 36 in. of a sprinkler head deflector.
d) Flammable and Combustible Liquids
   i. Approved safety cans and tanks must be used in handling and storing flammable and combustible liquids in quantities over one (1) gallon. Quantities of 1 gallon or less may be left in the original container or placed in an approved safety can.
   ii. Flammable/combustible liquids must not be stored in any area normally used for the safe passage of people.

iii. Flammable/combustible liquids in quantities over 25 gallons stored indoors must be placed in an approved safety cabinet. Not more than 60 gallons of flammable or combustible liquids shall be stored in any one safety cabinet. Follow manufacturer's limits.
iv. Flammable/combustible liquids stored outdoors in containers less than 60 gallons each must not exceed a combined amount of 1,100 gallons in one area or pile.
v. Each pile or area of storage of such flammable/combustible liquids stored outdoors must be separated by 5 ft. clearance and must not be within 20 ft. of a building or structure.
vi. Storage areas for flammable/combustible liquids must be kept free of weeds, debris and other combustibles not necessary to the storage.
vii. At least one (1) approved fire extinguisher must be located not less than 25 ft. and not more than 75 ft. from any outside flammable liquid storage area.
viii. At least one (1) approved fire extinguisher must be located within 75 ft. of every fueling area.
ix. There shall be no smoking or open flames within 50 ft. of fuel dispensing areas or other areas where fuel is used.
x. Engines of all equipment must be shut off during refueling operations.
xi. All portable fuel tanks (60 gallons or more) must be clearly labeled as to their contents and a "No Smoking" or "No Open Flame" sign placed on or near the tank.

15.8.3 Emergency Procedures (Fire)

a. Portable fire extinguishers control 85 to 90% of all construction fires. If a fire is encountered in its early stages, the nearest fire extinguisher should be used immediately. Simultaneously, others in the vicinity should be warned, and their help be solicited if appropriate. Operate the fire extinguisher as trained or as described on the instruction label. If the flames go out, do not continue to operate the extinguisher. Save its use as you stand by watching for a flare-up. Should the fire get out of control or be in an extremely hazardous area, leave the area while alerting others in the vicinity and call for outside emergency assistance. *(**Note:** Remember, water hoses, water buckets, dirt, blankets, etc. make excellent portable fire extinguishing agents.)*

b. **Call 911 or send someone to call 911** if a phone is available. Give the exact location of the incident, the type of emergency, and the type of equipment that may be needed.

NOTE TO MANUAL PREPARER: Consider the following paragraph only if the company regularly operates radios as part of the routine work force communications.

c. If a phone is not available, use the company radio to call base station. Give your unit number and say "Unit ____ to base—I have an emergency." This will let the base station operator and all other units know the urgency of the call. Repeat as needed. All other units are to remain silent, unless assistance is required of you. Give Base Station operator your exact location, type of emergency, and the type of equipment needed. Base Station operator or person receiving the radio transmission is to remind others as needed to "keep radio clear—emergency in progress." The Base Station operator will also give the "All clear—emergency over" notice.
d. Be certain that the company Safety and Health Manager is notified of the fire as soon as possible.

15.9 GASSES, VAPORS, FUMES, DUSTS, AND MISTS

a. Exposure to toxic gasses, vapors, fumes, dusts, and mists above acceptable concentration must be avoided.
b. When it is not feasible or possible to bring the area itself into compliance through the use of engineering and administrative controls, appropriate protective equipment or other protective measures are to be used to keep the exposure of employees to air contaminants within the limits prescribed. Any equipment and technical measures used for this purpose must be reviewed for each particular use by a technically qualified person. In such circumstances, employees will use all issued equipment at all times.

15.10 HAND TOOLS, POWER TOOLS, AND POWER-ACTIVATED TOOLS

15.10.1 General

a. No employee will operate electric, gas, or hand powered tools or equipment unless thoroughly familiar with use of the item and with all required safety precautions.
b. Faulty equipment shall be taken out of service immediately.

c. Employees will not use unsafe hand tools. Wrenches are not to be used when jaws are sprung to the point slippage occurs. Impact tools are to be kept free of mushroomed heads. Wooden tool handles are to be kept free of splinters or cracks, and are to be maintained tightly in the tool.
d. Employees are responsible to make a visual inspection of any tool before each use to ensure no unsafe conditions exist. If broken or excessively worn conditions exist, the tool must not be used until corrections or replacements are made.
e. Safety glasses or other approved eye protection must be worn by any employee using or being near a tool that normally or possibly could produce a projectile, splinters, or dust during its use or operation.
f. Manufacturer's switches and safety devices shall not be altered, removed, or overridden in any manner, including all types of guards, baffles, and shields.

15.10.2 Hand Tools

a. Impact tools such as hammers, chisels, wedges, drift pins, etc. must be kept free of mushroomed or splintered ends and must not be used until these unsafe conditions are corrected or the tool is replaced.
b. Tools with wooden handles that are loose, splintered, or cracked shall not be used until corrections or replacements are made.

15.10.3 Power, Power-Actuated, and Air Tools

a) Fuel Powered Tools
   i. All fuel powered tools must be shut off while being refueled, serviced, or maintained.
   ii. Smoking, matches, open lights or similar ignition sources are prohibited in the vicinity of any refueling operation.
   iii. Any fuel powered tool that develops a leak in the fuel system must be taken out of service immediately and repaired or replaced before continuing the operation.
   iv. When fuel powered tools must be operated indoors or in confined areas, or near occupied underground pits, manholes, vaults, etc. proper ventilation of such confined spaces must be maintained so as not to allow a hazardous atmosphere to form.

b) Electric
   i. All 110 volt power tools must either be approved double insulated or have a current assured grounding test I.D. (see Section 11.5). Those not meeting

one of these requirements must be taken to the shop for testing or replacement before use.

ii. Trailing cords of power tools are not to be used to lift or lower the tool.

iii. Bench and floor mounted grinders must have safety guards and work rests. Work rests must be kept within 1/8 in. from the wheel at all times.

iv. The use of any attachment wheel rated lower than the rated RPMs of the grinder is strictly prohibited.

c) Air Driven

i. Percussion type impact tools must have a safety clip or retainer to prevent the attachment from disengaging while in use.

ii. Compressed air in excess of 30 psi shall not be used for cleaning purposes. (*Exception:* concrete form cleaning and similar processes being performed by trained personnel.)

iii. All pressurized hose quick disconnect type (non-threaded) fittings, couplings, and connectors must have a safety mechanism providing a positive means to prevent accidental disconnect.

iv. All hose lines greater than 1/2 in. diameter are to be equipped with a safety device at the source or branch line from the source to automatically reduce pressure in case of hose failure.

v. Hoses are not to be used for lifting and lowering air tools.

d) Power Actuated Tools

i. Only persons trained in the proper use of the particular power actuated tool are permitted to operate it.

ii. The tool must never be pointed at another person whether it is empty or loaded.

iii. The tool must only be loaded just prior to its use and never left unattended when loaded.

iv. The tool must never be used near flammable or explosive materials or atmosphere.

v. Fasteners must never be driven into extremely hard or brittle materials, such as cast iron, glazed tile, face-hardened steel, glass block, face brick, etc.

vi. Fasteners must never be driven into materials that might be easily penetrated unless suitably protected against the possibility of rear exit.

e) Grinders—Abrasive

i. Abrasive wheel bench or stand grinders must have safety guards strong enough to withstand bursting wheels. Adjust work rests on grinders to a clearance not to exceed 1/8 in. between rest and wheel surface. Inspect and ring test abrasive wheels before mounting. Always leave wheel in working condition for next user. Properly dress wheel before using and/or when finished. The RPM rating of the abrasive wheel must be in accordance with the RPM rating of the motor.

f) Grinders—Portable (Air, Electric or Fuel)
   i. Portable grinders used with any wheel attachment must have guards, hoods, and/or safety flanges in place except when wheels less than 2 in. O.D. are used or when the wheel is entirely within the work being processed.
   ii. The use of any attachment wheel rated lower than the rated RPMs of the grinder is strictly prohibited.

15.11 HAZARD MATERIAL COMMUNICATIONS

15.11.1 General

a. Refer to Section 10 of this *Safety and Loss Control Program Manual*, **(COMPANY)** ***Hazard Communication Program,*** for complete description of the program.

15.11.2 The (COMPANY) Hazard Communication Program Manual

a. Refer to Section 10.2 of this *Safety and Loss Control Program Manual* for description of the use of the **(COMPANY)** ***Hazard Communication Program Manual***.

15.12 HEAVY EQUIPMENT AND TRUCKS

15.12.1 General

a. Manufacturer's Operator Manuals and safety information shall be readily available to the operator at all times.
b. Any special load rating capacity charts, recommended operating speeds, hazard warnings, or similar instructions shall be posted where they are visible to the operator while at the control station.
c. An appropriate fire extinguisher in good operating condition, properly sealed, with a current inspection certification, shall be kept with all motorized heavy equipment and trucks.
d. All cab glass must be safety glass or the equivalent and must be kept clear of

visible distractions affecting safe operation. All broken or cracked glass must be replaced.

e. Seat belts must be provided and used on all heavy equipment and trucks with the exception of equipment not requiring rollover protection, or that is designed for stand-up operation only.
f. It is the operator's responsibility to see that no personnel are allowed to ride on their equipment or truck unless there are seats installed for this purpose and the person is seated properly. Truck beds are *not* to be used to haul personnel under any circumstances.
g. All heavy equipment and trucks must have a functional horn at the operator's station.
h. All heavy equipment and trucks shall be equipped with an automatically operated back-up alarm that can be heard above the surrounding noise level. Exceptions are vehicles used only for on-road driving, and vehicles 1 ton or less that have a clear view to the rear.
i. In a temporary situation, a vehicle with a recently inoperable alarm can be backed if the operator has a clear view to the rear or if an observer is used to assure the operator that all is clear to the rear.
j. All equipment and trucks that are equipped with apparatus that can be raised and lowered from the operator's station (i.e., backhoes, endloaders, dump trucks, forklifts, etc.) must be placed in the "grounded," closed, or lowered position before the operator leaves the control station. This would also apply when the operator has completed one task and will be waiting a substantial or unknown amount of time until the next task.
k. Operating levers that control dumping/hoisting devices on dump trucks must have a latching or other device that prevents accidental operation.

15.12.2 Maintenance and Repairs

a. During the maintenance or repairs of any vehicle or piece of equipment, the person(s) doing the work must first take steps to assure that no one will inadvertently attempt to operate it (i.e., tag-out, lock-out, disable power source, remove keys, etc.).
b. During the maintenance or repairs of a piece of equipment or truck which must be performed with apparatus in the opened, raised, or lifted position, the apparatus must be physically blocked, pinned, locked, or otherwise supported to prevent accidental lowering or closure of the apparatus.
c. Should it become necessary for a person performing repairs or maintenance on a piece of equipment or vehicle to leave even for a short time prior to completing the work, the steps in 15.12.2.a above must be applied.

15.12.3 Inspections

a. The operator or a designated competent person must make a safety inspection of the equipment to be used prior to each work shift. Any conditions found that can result in unsafe operation must be corrected before continuing. This applies to conditions arising during operation as well. Refer to Section 15.5 for inspections to hoisting equipment.

15.12.4 Site Clearing Equipment

a. All equipment used in site clearing must have an overhead canopy of at least 1/8" steel plate or 1/4" woven wire mesh of a maximum 1" openings. Openings or windows in the rear of the canopy must have 1/4" woven wire mesh of maximum 1" openings or equivalent protection.

15.12.5 Cranes, Draglines, and Hoisting Equipment

a. All hand signals to cranes, etc. must be those prescribed by the American National Standards Institute (ANSI) for the specific crane in use.
b. An annual inspection must be made by a competent person or agency and records maintained for each piece of hoisting equipment.
c. The operator or a designated competent person must periodically check all wire rope used in conjunction with hoisting equipment. Wire rope found to be out of compliance with OSHA 1926.550(a)(7) must be taken out of service.
d. The operator shall be responsible to see that no person is permitted to be lifted, lowered, or ride on any hoisting equipment's load lines, attachments to load lines, boom, etc.
e. The operator shall never lift a load over the head of another person, including persons in vehicles, equipment, trailers, etc.
f. The operator shall consider all signaling not understood as a "Stop" signal.
g. The operator shall be responsible to see that the rear swing radius of a crane or dragline that is accessible to personnel is barricaded to prevent persons from becoming struck or crushed by the crane's rear structure.
h. Overhead cranes shall be operated only by authorized persons that have received instructions on their proper use.
i. When the operator does not have a clear unobstructed view of the hoisting operation, intermediate signaling person(s) or radios must be used.

15.12.6 Procedure to Remove Rebar or Other Objects from Equipment Tracks, Tires, or Undercarriage

a. If concrete reinforcing steel (rebar) or any other object becomes lodged in any equipment track, tire, or undercarriage immediately:
   i. ***STOP THE EQUIPMENT***. Secure it in park and shut off the engine.
   ii. Notify your immediate supervisor.
   iii. Notify, or ensure that your supervisor notifies, the (**COMPANY**) Safety and Health Manager.
   iv. ***DO NOT*** resume operation of the equipment until:
      a) Specific instruction has been given by your supervisor, and
      b) The corrective procedure is provided under the guidance of the Safety and Health Manager and/or the company equipment foreman.

15.13 HOISTS—MATERIAL AND PERSONNEL

a. Rated load capacities, recommended operating speeds, and special hazard warnings or instructions posted on cars and platforms may not be exceeded under any circumstances.
b. (Material hoistway entrances hoists) will be protected by substantial full width gates or bars.
c. Hoistway doors or gates of personnel hoists will be not less than 6 ft. 6 in. high, and will be protected with mechanical locks which cannot be operated from the landing side and are accessible only to persons on the car.
d. No riders are allowed on material hoists except for inspection or maintenance.

15.14 HOUSEKEEPING AND DISPOSAL OF WASTE MATERIAL

15.14.1 Housekeeping

a. Generally, housekeeping responsibilities fall on every employee to see that their immediate work area is free from housekeeping hazards that might cause slips, trips, falls, fires, health hazards, electrical hazards, etc. Similarly, each employee should be looking out for similar hazards while walking from place to place during their daily work activities, and should correct or report any such hazards at once.

b. Jobsites must be maintained in such a manner that no debris is allowed to accumulate in or to clutter work areas, walkways, stairs, ladders, aisles, doorways, trailer floors, etc.
c. Form and scrap lumber with protruding nails and all other debris will be kept clear from work areas. Remove combustible scrap and debris at regular intervals.
d. Depending upon the specific requirements of the individual subcontracts, containers may be provided for collection and separation of all refuse.
e. Covers are required on containers used for flammable or harmful substances.
f. No glass containers allowed on jobsites.
g. At the end of each portion of work, return all tools and excess material to proper storage. Each employee is responsible for clean-up of all debris from their own work areas before moving on to the next phase of work.

15.14.2 Disposal of Waste Material

a. Combustible scrap materials shall be removed from jobsites at intervals sufficient to prevent unnecessary fire potential.
b. Scrap or form lumber with protruding nails must be processed while being disassembled to eliminate the obvious hazard of the sharp nail points. Removal of the nail is desired, however, pounding the point back into the board or bending it over is acceptable when the lumber is to be disposed of shortly.
c. Garbage that might present a health hazard, such as decomposing food, acids, harmful dusts, etc. shall be kept in covered containers and disposed of at frequent intervals.
d. All flammable waste such as certain solvents, oily rags, etc. must be kept in fire resistive covered containers until removed and disposed of.
e. All materials classified as hazardous must be disposed of using approved methods described in the manufacturer's MSDS (Material Safety Data Sheet) or other legally approved information.
f. All disposal of waste material and debris by burning shall comply with local fire regulations. Burning permits must be obtained where required. Changing wind conditions must be considered and action taken to prevent any uncontrolled burning. At least one person must be in the general area during the burn to monitor conditions until the fire is completely out.

15.15 LADDERS, STAIRWAYS, AND RAILINGS

15.15.1 Ladders

a. Always face the ladder when moving up or down on it, keeping at least one hand grasping the ladder.
b. Never carry a load on a ladder that might cause you to loose your balance or fall.
c. Never use the top step of a ladder as a step.
d. All ladders must be kept free of oil, grease, and other slipping hazards.
e. Any ladder that becomes defective, weakened, broken, or otherwise faulty, must be tagged "DO NOT USE" and placed out of service until repaired.
f. Ladders used where they could possibly come in contact with energized electrical equipment or lines must have non-conductive side rails.
g. Ladders used to access an elevated area or landing must extend at least 3 ft. above the landing. If not possible, then the ladder must be secured at the top and a grab rail or similar grasping device at the landing be provided.
h. Ladders that are not permanently fixed or self-supporting shall be used at a ratio of approximately 4:1 (i.e., ladder being used at a *working length* of 16 ft. would need to be 4 ft. from the wall at the base).
i. Ladders used on unstable or unlevel surfaces must be secured to prevent their displacement.
j. Ladders used in unbarricaded areas of traffic such as doorways, driveways, etc. must be secured against displacement. Ladders used on hard, smooth, slippery type surfaces must be secured or must have slip-resistant safety feet.
k. Fixed ladders must be constructed so that a minimum clearance of 7 in. is maintained behind the ladder and 30 in. is maintained at the front (climbing) side of the ladder.
l. Job-made ladders can be constructed and used with prior approval from the Safety and Health Manager. Requirements of ANSI Standard A14.4 must be met in their construction and use.

15.15.2 Stairways

a. Generally any personnel access area with a difference in elevation of 9 in. or more must have an approved means of moving from one elevation to the other, including stairways, ramps, sloped embankment, hoist, etc.
b. Stairways having four or more risers or raising more than 30 in. must have a handrail and midrail along each unprotected side. The handrail shall be between 30 and 37 in. high.

c. Where doors or gates swing open into a stairway, a landing platform or area must be maintained that is a minimum of 20 in. in width.

15.15.3 Railings

a. A standard railing will consist of top rail, intermediate rail, toeboard and posts, and have a vertical height of approximately 42 in. from upper surface of top rail to floor, platform, etc. The top rail of the railing will be smooth surfaced, with a strength to withstand at least 200 pounds. The intermediate rail will be approximately halfway between top rail and floor.
b. A stair railing will be of a construction similar to a standard railing, but the vertical height will be no more than 34 in. or less than 30 in. from upper surface of top rail to surface of tread in line with face of the riser at the forward edge of the tread.

15.16 LASERS

a. Only qualified and trained employees will be allowed to operate lasers. Employees will wear proper eye protection where there is a potential exposure to laser light greater than 0.005 watts (5 milliwatts).
b. Proof of qualification of laser operator must be available at all times.
c. Beam shutters or caps will be utilized, or laser turned off when laser transmission is not actually required. When lasers are left unattended for a substantial period of time, they are to be turned off.

15.17 MASONRY WALL ACCESS ZONE

a. Limited access zones are to be established on the unscaffolded side of unbraced masonry walls. The zone is to be equal to the finished height of the wall plus 4 ft.

15.18 MATERIAL STORAGE, HANDLING, AND RIGGING

15.18.1 Storage

a. Material shall not be stored in aisles, passageways, doorways, etc., nor shall it obstruct or adversely affect any fire-fighting equipment.

b. No material may be stored within 36 in. of a sprinkler head deflector.
c. Bagged materials shall be stacked by stepping back the layers and cross keying the bags at least every 10 bags high.
d. Brick stacks shall be limited to 7 ft. maximum height. Loose stacks shall be stepped back 2 in. per foot in height for every foot over 4 ft.
e. Masonry block shall be stepped back 1/2 block per tier above the 6 ft. high level. Lumber must be stacked free of nails, level, self-supporting and a maximum of 16 ft. high.
f. Poles, pipes, bars, etc. shall either be stored in racks or stacked and blocked to prevent rolling or spreading.
g. Cans, drums, pails, etc. to be stored loose must be stepped back 1 container per tier above the 4 ft. high level.

15.18.2 Handling and Rigging

a. Rigging equipment shall be inspected prior to use on each shift and as needed during its use to ensure it is safe. Defective or suspicious equipment must be removed from service immediately.
b. Makeshift rigging equipment is prohibited.
c. Use of wire rope:
   i. Safe working loads for wire rope rigging equipment, as provided by the manufacturer or as determined by approved informational charts (i.e., OSHA 1926.251 Tables H-3 through H-14), shall not be exceeded when handling material, etc.
   ii. Wire rope used in lifting, lowering or pulling loads shall be one continuous piece without splice.
   iii. Eye splices in lifting bridles and slings shall not be made with wire rope clips or knots.
   iv. When using wire rope clips or similar items to form eyes and splices in cables, the manufacturer's instruction must be strictly followed with respect to size, number, and installation procedure of the particular clips or similar items.
   v. Wire rope rigging equipment shall be taken out of service when in any length of 8 times the diameter, the total number of visible broken wires exceeds 10% of the total wire construction of the wire rope (i.e., a 6 × 7 constructed 1" diameter cable has 42 wires. 10% of 42 = 4 wires. 1" diameter × 8 = 8 ins.; thus, if the cable has more than 4 visible wires broken in any 8 in. section it must be replaced).
d. All shackles and hooks used in rigging shall exceed the manufacturer's safe working load rate for the load being handled. Unidentified shackles must conform to an approved safe workload informational chart (i.e., OSHA

1926.251 Table H-19). Unidentified hooks must be treated and documented at (2) times the intended load before using.

15.19 OPENINGS: FLOORS, WALLS, ELEVATOR SHAFTS, AND SIMILAR AREAS

a. All openings in floors, walls, elevator shafts, and similar areas are to be protected with a properly designed guard rail and toeboard, or are to be covered in a manner that will prevent falls. Provide railing on all exposed sides, except at entrances to stairways.
b. Every open-sided floor or platform that is 6 ft. or more above the adjacent floor or ground level must be protected by a properly designed railing, or equivalent, on all open sides except where there is an entrance to a ramp, stairway, or fixed ladder.
c. Ladderway floor openings or platforms must be protected with appropriately designed guard rails and standard toeboards on all exposed sides, except at entrances to openings. Passage through the railing is to be provided by a swinging gate, or must be offset so that a person cannot walk directly into the opening.
d. Runways 4 ft. high or more need standard railings on all open sides.
e. Floor holes into which persons can accidentally walk are to be protected by either an appropriately designed railing with toeboard on all exposed sides, or an appropriate floor hole cover. While the cover is not in place, the floor hole will be protected by a properly designed railing.
f. Wall openings through which there is a drop of more than 4 ft., and the bottom of which is less than 3 ft. above the working surface, must be protected. When the height and placement of the opening in relation to the working surface is such that a properly designed rail will effectively reduce the danger of falling, the rail must be provided.
g. The bottom of a wall opening which is less than 4 ft. above the working surface will be protected by a properly designed toeboard or an enclosing screen.

15.20 PERSONAL PROTECTIVE EQUIPMENT

15.20.1 General

a. In general, the employee is responsible for wearing and using appropriate personal protective equipment in operations where there is exposure to hazardous conditions or where need is indicated to reduce hazard. The company will always endeavor to anticipate all such needs and issue appropriate equipment, but it remains the employee's responsibility to use all such issued equipment, and to advise company management if needed equipment had not been issued.

15.20.2 Proper Clothing (Field Personnel, Shop Personnel, and Truck Drivers)

a. General: All clothing shall be adequate to protect the employee from the known and suspected hazards normally acquainted with the type of work to be performed.
b. Shirts:
   i. Must be worn at all times.
   ii. Must not have large rips, tears, holes, etc.
   iii. Must extend to at least the waistline.
   iv. Must have full top of shoulders protection (i.e., tank tops are *inadequate*, muscle T-shirts are *adequate*).
   v. Loose, floppy, etc. shirts shall not be worn around equipment that could catch the material and pull the person into it.
c. Pants:
   i. Must be full length.
   ii. Must not have large rips, tears, holes, etc.
d. Footwear:
   i. Must be of leather or solid man-made materials; not canvas, rubber, etc. (i.e., no tennis or running shoes).
   ii. Must not have large rips, tears, holes, etc.
   iii. The employee must wear socks that protect both foot and ankle.
   iv. A good heavy work boot that at least covers your ankle is highly recommended. Steel toe and steel shank boots provide even better protection and are strongly encouraged.

15.20.3 Head Protection

a. Approved hard hats in good repair and with proper suspension are required to be worn by all employees at all times on every project, when not under separate overhead protection (i.e., inside a nearly finished building, inside equipment or vehicle with a roof, etc.).
b. Employees working in or around the main office/shop complex will only be required to wear hard hats in areas where there is a potential danger of head injury due to falling, flying, or moving objects.

15.20.4 Hearing Protection

All hearing protection used by any employee:

a. Must be approved, of known attenuation type supplied by the company, or approved by the company Safety and Health Manager.

b. Must be used as described by manufacturer's directions and/or company designated instructors.

c. Must be used in any work area where the noise level is found to be above the permissible levels of Table 1 below, after all feasible engineering and administrative controls have been implemented in effort to reduce the noise conditions of the respective work area(s).

Table 1. **Permissible Noise Levels**

Decibels of Noise (dbs)	**Worktime Allowed Per Day**
90	8 Hours
92	6 Hours
95	4 Hours
97	3 Hours
100	2 Hours
102	1.5 Hours
105	1 Hour
110	.5 Hour

d. Portable radios, tape players, and similar devices that require earphones or ear plugs are strictly prohibited while working (break and lunch times excluded). This restriction especially applies while driving a company vehicle or operating any equipment.

15.20.5 Eye and Face Protection

a. Protecting eyesight with proper eye protection full time during the work day is highly recommended.

b. Approved industrial safety glasses and/or eye protection shall be worn by employees while performing any duty or working near any operation where the potential exists for an eye injury from physical, chemical or radiant agent.
c. Employees who must wear corrective lenses that are not of an approved industrial safety glasses type are required to wear safety goggles over their personal corrective lenses to comply with (a) above.
d. Employees who are working with or near a laser beam operation must never look into or allow the beam to contact their eye(s). Normal safety glasses will not provide adequate protection against radiation damage to the eye, if exposed.
e. All welding, torch and gas soldering operations require approved eye protection including a shaded lens. The following Table 2 is a guideline to be used in selecting the correct filtered lens shade number:

Table 2. **Filger Lens Shade Numbers**

Operation	**Shade Number**
All forms of arc welding	10–14
All forms of gas welding	4–8
Braxing and cutting	3–5
Gas soldering	2

15.20.6 Respiratory Protection

a. No employee shall enter or work in an area where a known or suspected toxic or harmful atmosphere may prevail. The Safety and Health Manager must be notified to allow for proper testing of the area and for prescribing the proper respiratory protection program or other safeguards to be initiated. This includes areas of potential low levels of oxygen.
b. In all situations in which employees are working in areas where abnormal levels of common nuisance dusts, fumes, or mists prevail, an approved dust mask must be worn.
c. In emergencies, or when feasible engineering or administrative controls are not completely effective in controlling toxic substances, approved respiratory protective equipment must be used by every employee working in such conditions. Respiratory protective devices will be approved and issued for the

hazardous material involved and extent and nature of work requirements and conditions.

d. Employees required to use respiratory protective devices will be thoroughly trained in their use. Respiratory protective equipment will be inspected regularly and maintained in good condition.

15.20.7 Safety Belts and Lifelines

a. An approved safety belt with maximum 6 ft. lanyard attached to a structure able to hold 5,400 lbs. shall be worn by any employee working in an area in which a fall of 10 ft. or more is possible, or in which a fall into a hazardous area is possible (such as into a tank or vat of chemical, into a working part of a machine, and into the path of moving vehicles).

b. A lifeline able to hold 5,400 lbs. attached to a structure of comparable strength must be used to support a safety belt, where required.

15.20.8 Life Jackets and Ring Buoys

a. Employees working over or near water where a life threatening hazard exists must wear a life jacket at all times.

b. A ring buoy with at least 90 ft. of throw line shall be readily availabie also to supplement (a) above.

15.20.9 Back Support Belts

a. A mandatory Back Belt Program is in effect for all field and shop personnel. This program applied to all mechanics, truck drivers, equipment operators, and field personnel at “working foreman” level and below. Supervisors, managers, office personnel, etc. will be exempt from the mandatory requirement, but are advised to use the back belts whenever engaged in any movement or lifting operations that carry the possibility of back injury.

b. The belts issued are equipped with suspenders, are made of breathable materials, come in various sizes, and are adjustable. They are to be worn throughout the workday. They should be adjusted loose when not doing strenuous activities and snug when engaged in work that strains the lower back. Failure to wear and use your back belt will be treated as any other violation of job safety rules and regulations. An individually signed receipt will be used to document each employee being issued the belt and the employee’s understanding of replacement costs, should the belt be lost or abused. Normal wear and tear or replacements of faulty material will be made at the company's expense. Proper care, including security from theft and washing as needed, will be the issued employees' responsibility. Exceptions to full time mandatory back belt program are the following:

i. During normal break time, lunch time or while riding in vehicles.
ii. Temporarily, as approved by employee's foreman or supervisor, when the possibility exists that the belt or suspenders could get caught or snagged on equipment or structure members causing a hindrance or potential hazard to the employee. The belt must be put back on as soon as the immediate task is completed.

15.21 Safety Nets

a. Safety nets are required when workplaces are more than 25 ft. above the surface and the use of ladders, scaffolds, catch platforms, temporary floors, safety lines, or safety belts is impractical.

15.22 SANITATION CONTROLS

15.22.1 Drinking Water

a. An adequate supply of fresh drinking water is to be readily available at every jobsite.
b. Where portable water containers are used, they must be kept clean, tightly closed, never dipped into for ice or water, and never used as a cold storage for other items such as lunches, soda, etc.
c. Each portable drinking water area shall be supplied with single use disposable cups, a clean container to hold cups, and a cup disposal container.

15.22.2 Toilet Facilities

a. Where permanent toilet facilities are not readily available to workers at a jobsite, portable facilities of the chemical type will be provided as required by the particular agreement for construction services.
b. Under temporary field conditions or on jobsites with one to twenty employees, there must be at least one seated toilet.
c. Twenty or more employees at a jobsite must be supplied with a minimum of one seated toilet and one urinal per forty workers.

COMPANY LOGO	Construction Safety and Loss Control Program Manual	Page: 15-34	Appr:
	15. SAFETY AND LOSS CONTROL ON THE JOBSITE	Revision Date:	

15.23 SCAFFOLDS

15.23.1 General Scaffolds

a. All scaffolds will be capable of supporting 4 times maximum their intended load and erected on sound, rigid footing, capable of carrying the maximum intended load without settling or displacement.
b. Guard rails and toeboards will be installed on all open sides and ends of platforms more than 6 ft. above ground or floor, except needle beam scaffolds and floats which require the use of safety belts. Scaffolds 4 ft. to 10 ft. in height, with a minimum dimension in either direction of less than 45 in., will have appropriate guard rails installed on all sides and ends.
c. Where persons are required to work or pass under scaffolds, there will be a screen with maximum 1/2" openings between toeboard and guard rail. Planking will be scaffold grade or equivalent as recognized by approved grading rules for the species of wood used in the planking. Scaffold planking is to be overlapped a minimum of 12 in. or otherwise be secured from movement.
d. Scaffold planks will extend over end supports not less than 6 in. nor more than 12 in. Scaffolding and accessories with defective parts will be immediately repaired or replaced.

15.23.2 Mobile Scaffolds

a. All platforms are to be tightly planked for the full width of the scaffold, except as may be necessary for entrance openings. Platforms will be secured in place.
b. Guard rails made of lumber are not to be less than 2×4 in. (or equivalent), approximately 42 in. high, with a midrail of 1×6" lumber (or equivalent). Toeboards will be installed at all open sides and ends on scaffolds more than 10 ft. above ground or floor. Toeboards will be minimum 4 in. in height. Where persons are required to work or pass under scaffolds, wire mesh is to be installed between toeboard and guard rail.

15.23.3 Swinging Scaffolds

a. On suspension scaffolds designed for a working load of 500 pounds, no more than two workers are permitted to work on the scaffold simultaneously. On suspension scaffolds with a working load of 750 pounds, no more than three workers are permitted on the scaffold simultaneously. Each employee on the scaffold must use an approved safety life belt attached to a lifeline. The lifeline is to be securely attached to substantial members of the structure (not the scaffold itself) or to securely rigged lines which will safely suspend a worker in event of a fall.

15.23.4 Tubular Welded Frame Scaffolds

a. Scaffolds will be properly braced as recommended by the manufacturer by cross bracing or diagonal braces, or both, for securing vertical members together laterally. Cross braces will be of such length as will automatically square and align vertical members or erected scaffold is plumb, square, and rigid.

15.24 SECURITY AND VANDALISM

a. The planning, layout, and monitoring that goes into a jobsite security and vandalism control program is usually the difference between loss control and expensive thefts and damage. A jobsite without guards, fencing, adequate lighting, or controlled exits presents an easy target for even the inexperienced vandal or thief.
b. Before or during the initial project mobilization, the project supervisory staff should meet with the local police, state highway patrol, or county sheriff and the local fire department. The law enforcement professionals should be provided with the details of the project, the type of construction, anticipated work schedules, planned starting times and scheduled dates of completion. The names of key personnel, telephone numbers, and how to reach them during non-working hours should also be given. The authorities should be kept informed at all times. Request a crime prevention survey of your site, and include them to the greatest extent possible in all pre-job security planning.

15.25 SIGNS, SIGNALS AND BARRICADES

15.25.1 Signs and Barricades

a. In cases where employees are required to work on or adjacent to a public roadway, proper signs and barricades warning and directing the area traffic must be in place prior to work proceeding and must be maintained and modified as needed throughout the time frame of the activity. Whenever possible, the arrangement should be accomplished through a contractual agreement with an outside company as determined by a particular project's needs. Such an arrangement, however, does not eliminate the responsibility of the project manager to assure that adequate signage and barricades are being properly used and appropriately maintained.

b. Proper signs and barricades are to be used on (**COMPANY**) jobsites whenever needed, in order to warn and direct pedestrian traffic and our own employees walking and working in the vicinity of a potential hazard, such as an open manhole, trench, falling or flying debris, high pressure pipe testing, etc.

15.25.2 Signaling (Traffic Control)

a. When operations on or adjacent to a public roadway are such that signs and barricades do not adequately protect employees or warn and direct the traffic, flagmen or local law enforcement must be used.
b. Flagmen and any other employee working on or near a public roadway must wear a brightly colored orange or red safety vest. The vest must have reflective material incorporated front and back if used at night or in any dim-light conditions.
c. Signaling by flagmen should be done by the use of brightly colored red or orange flags 18 in. square (minimum) or by use of proper sign paddles or pivot posts.
d. Flagmen working in conjunction with one another must maintain good visual or other appropriate communications at all times.
e. Signals used by flagmen must conform to those established under the ANSI Manual or Uniform Traffic Control Devices. Consult with the company Safety and Health Manager for additional assistance and detail.

15.25.3 Guarding Opened Manholes and Similar Structures

a. A single crew must not have more than two (2) manholes opened at any one time. *Exception:* Three (3) may be opened for short durations during outside agency and client inspections. An extra manhole guard should be posted if needed.
b. Opened manholes must be barricaded with free standing metal manhole guards, cones with flagging encompassing entire manhole, or stakes and flagging encompassing manhole. Metal manhole guards must have chains in place when left unattended. Refer to (c), below.

c. An opened manhole not properly barricaded for any reasons (refer to (b), above) must be attended at all times. A manhole is considered unattended if there is no employee above ground within arm's length of the opened hole keeping persons away from it.

d. Where several manholes on a jobsite must be accessed periodically throughout a work day, temporary plywood manhole lids may be used to cover those manholes not otherwise barricaded.

15.25.4 Accident Prevention Signs and Tags

a. Signs and symbols required by hazardous jobsite conditions are to be visible at all times when work is being performed, and are to be removed or covered promptly when the hazards no longer exist.

b. Types of signs include:
   1) **Danger Signs**
      a. Are to be used where immediate hazards exist.
      b. Shall have red as the predominant color for the upper panel; black outline on the borders; and white lower panel for additional sign wording.
   2) **Caution Signs**
      a. Are to be used to warn of potential hazards, or to caution against unsafe practices.
      b. Shall have yellow as the predominant color; black upper panel and borders; yellow lettering of "caution" on the black panel, and lower panel for additional sign wording. Black lettering shall be used for additional wording.
   3) **Exit Signs**
      a. When required, shall be in legible red letters not less than 6" high on a white field.
   4) **Safety Instruction Signs**
      a. When used shall be white with green upper panel with white letters to convey the principle message. Any additional wording shall be black letters on the white background.
   5) **Directional Signs**
      a. Other than automotive traffic signs described in (6) below, shall be white with a black panel and white directional symbol. Additional wording on the sign shall be black letters on a white background.
   6) **Traffic Signs**
      a. Construction areas are to be posted with legible traffic signs at points of hazard.

b. All traffic control signs are to conform to American National Standards Institute D6.1–1971 Manual of Uniform Traffic Control Devices for Streets and Highways.
c. Refer to the company Safety and Health Manager for additional direction or confirmation of proper signage procedure.

15.26 TAG-OUT AND LOCK-OUT PROGRAMS

15.26.1 Safety Tag—"Out of Service"

a. This tag is authorized to be used by the company Safety and Health Manger or his/her designee only. The tag will be used to tag "out of service" on any type of equipment, machinery, tools, etc. deemed necessary.
b. The tag is in black letters on a red background and has the universal danger sign on one side.

15.26.2 General Tag-Out/Lock-Out Program

a. This program is being implemented in order to provide additional protection to employees who may need to work on or near potentially hazardous electrical or mechanical equipment that has been deenergized, disconnected, or otherwise disabled. The person responsible for the deenergizing process will also be responsible for assuring that the equipment is tagged out of service and, wherever possible, locked out. The tag must be signed and dated and must be hung on or within 12 in. of the switch, disconnect, breaker, etc. that is shut off.
b. Upon completion of work, the person who tagged out/locked out the equipment will be responsible for removal of all tags and locks. If that person is not available at the work site, then his or her immediate supervisor may designate a qualified person to perform this task, provided the supervisor informs the original "tag–out-of-service" person of such action before he or she resumes work at that work site. In either event, the responsible person must make a visual inspection to clear the affected area and equipment prior to reenergizing it.
c. Tags and locks used to comply with this program can be obtained from the company Safety and Health Manager. Ongoing company safety training classes will periodically include instructions concerning this program.

Should there be any questions about the Tag—Out of Service / Lock Out program, please contact the Safety and Health Manager.

d. The tag is black letters on white background

15.27 TRENCHING AND EXCAVATING

15.27.1 General Requirements

a. Preplanning of every excavation by the responsible supervisor is required in order to prevent problems with items such as underground utilities, vehicular and pedestrian traffic, overhead utility lines, surface drainage, substructures and superstructures, excessive noise laws, environmental impact laws, medical and/or emergency response facilities, potential for hazardous atmosphere, communications and back-up availability, and normal amounts of vibration in the area.
b. Dewatering as required to sustain a dry excavation must be initiated and maintained while work is being performed.
c. Persons working in or around an excavation must be alert to any unnatural or suspicious changes in the excavation. Any observation or concern must be reported to their immediate supervisor immediately. Such concerns may include cracks in walls or surface, water seepage, suspicious odors, change in surrounding vibrations, incoming drainage, etc.
d. Prior to starting any work in the excavation area, the immediate supervisor (designated Competent Person) shall inspect the excavation for possible hazards and to see that all safety regulations are met. Additional inspections will be made throughout the day as required by changing conditions and at the beginning of each new work day or work shift. Each inspection is to be documented by the designated Competent Person using the appropriate company form, or as may be adequate on the project's complete daily field report. Any separately completed forms must be turned in to the Safety Department on a regular basis.
e. The immediate supervisor (Competent Person) must remain in the general vicinity of any excavation in which persons are working. Should the designated Competent Person find it necessary to leave, they must designate a qualified person as Competent Person. This designee must have a working knowledge of the trenching and excavation regulation, be able to identify existing and potential hazards associated with the operation, and have authority to take immediate corrective measures. Should no such person be available, work within the excavation must be halted until the Competent Person returns.

f. At the end of the work shift or when the work area must be left unattended at any time, the immediate supervisor shall perform a walk-around inspection of the area assuring that no hazardous or unsafe conditions are left exposed or unprotected.

g. **(COMPANY)** policy requires that all field employees wear a safety vest at all times when performing any activities on any jobsite not inside a building structure. In any event and at a minimum, employees working on or near a public roadway must wear a safety vest. Refer to section 15.25.2b for requirements.

h. Where a condition exists that vehicular traffic poses a potential hazard to persons working in or around an excavation, adequate barricades, signage, earthen berms, signalmen, barriers, etc. must be provided that would best control the hazard.

i. The Safety and Health Manager must be informed of any proposed excavation in excess of 20 ft. deep in order to allow for the proper engineering or optional protection system(s) to be planned and implemented.

15.27.2 Excavation Access and Egress

a. All excavations 4 ft. or deeper must have a ladder, ramp (earthen or structural), or similar safe means of access/egress within 25 ft. of persons working in the excavation.

b. Ladders used in excavations as access/egress must be secure and must extend at least 3 ft. above the top landing area or have an adequate grab rail or grab bar system at the top.

c. Earthen ramps must be at no greater incline than that which would allow a person to exit in an upright position.

d. Structural ramps must have a non-slip walking surface or must be supplied with cleats.

e. Employees entering bell-bottom shaped excavations (i.e., pier holes) or similar deep, confined footing excavations must wear a safety harness with a lifeline at all times. The lifeline must be continually attended by person outside the excavation.

15.27.3 Hazardous Atmospheres

Refer to Section 9.3, "Confined Area Entry" of this *Safety and Loss Control Program Manual*, for appropriate instructions.

15.27.4 Soils Classifications

NOTE TO MANUAL PREPARER: The following sample sections apply to contractors who are engaged in site utility and site development work. You may consider deleting the sitework-specific material from your own Safety and Loss Control Program Manual if such work is not performed by your company, or you may decide to leave the material in, in order to present the most comprehensive Safety and Loss Control Program Manual possible.

a) There are basically (4) types of soil classifications used to determine stability of soils:

Stable Rock
Type A (cohesive)
Type B (mildly cohesive)
Type C (non-cohesive)

When in doubt as to the soil type in a particular excavation **Type C** is to be assumed for the purpose of safety in trenching and excavating. All sloping, benchings, and protection systems used on our projects must conform to Type C soil requirements. Exceptions will only be allowed when authorized and classified by the Safety Department.

b) The immediate supervisor (Competent Person) must make at least one visual test and one physical or manual test to determine the soil classification.

1) **Visual**
By looking at the soil around the excavation, the soil that is excavated, or the walls of the excavation and deciding whether it is:

Cohesive A-Type
(high clay or cemented soil content)

Mildly Cohesive B-Type
(high loam or silt mixed with clay or sand)

Non-Cohesive C-Type
(sandy, granular soil or extremely wet soil or clay).

2) **Physical**
By seeing how difficult it is to press your thumb into an undisturbed section or clump of soil and how deep a penetration is obtained

Cohesive A-Type
(hard pressure with little to no penetration)

Mildly Cohesive B-Type
(mild pressure and full penetration)

Non-Cohesive C-Type
(easily penetrated several inches).

Remember: Work in A and B soil classifications must be authorized by the Safety and Health Manager.

15.27.5 Sloping

a. All excavated materials, spoils, loose dirt, stones, tools, etc. must be kept back at least 2 ft. from the edge of an excavation or adequate retaining shields or blocking devices must be used to ensure that materials cannot fall onto employees in the excavation.
b. Type-C soils must be sloped to an incline of at least 1:1; that is, for every 1 ft. depth you must have top of slope back 1 ft. from toe of slope or 1 ft. horizontal for every 1 ft. vertical.

15.27.6 Benching

a. Type-C soils—Generally benching is not permitted in Type-C soils unless used in combination with a trench box or shoring. Exceptions are made for layered soils where Type-C soils are above Type-B or Type-A soils. In such conditions, the Safety and Health Manager or company official must be notified for authorization.
b. Benching is permitted in Type-C soils for any depth excavation provided that the upper bench or benches are excavated back far enough from the edge of the next lower top of slope to virtually make the upper bench non-affecting the lower bench; for example, scalping an area 4 ft. deep and 50 ft. wide for a proposed trench 10 ft. deep and 10 ft. wide. The result would be a trench 6 ft. deep so that the closest bench to the top of its slopes would be back 20 ft., which would allow an 8 ft. high trench box to be used safely.

15.27.7 Shoring

a. Any excavations deemed necessary to use shoring will be handled on an individual basis. The proper materials or system to be used will be determined through a combined effort of the Competent Person, field supervision and the Safety and Health Manager.

b. Generally, shoring systems may include the following:
   i. ***Designs for timber shoring in trenches*** shall be determined by the Competent Person or registered professional engineer. For aluminum hydraulic shoring, designs will strictly meet or exceed requirements expressed in the manufacturer's tabulated data.
   ii. ***Designs using manufacturers' tabulated data*** may be used; deviation is not allowed.
   iii. ***Designs using other tabulated data*** may be used provided the data is in writing and includes: Explanatory information to aid the user in making a selection, the criteria determining the selection, and the limits on the use of the data. At least one copy of the information, including the identity of the registered professional engineer, must be kept at the worksite during construction of the protective system.
   iv. ***Design by a registered professional engineer.*** Design systems not using any of the three previously cited options must be approved by a registered professional engineer, shall be in writing, and include the identity of the registered professional engineer and details such as sizes, types, and configurations of the materials to be used. At least one copy of the plan is to be at the jobsite during construction and throughout the period of work in the shored excavation.

15.27.8 Trench Boxes (Shields)

a. Trench boxes may be used in excavations where sloping requirements cannot be met or are an undesirable option for protection against cave-ins.
b. Only trench boxes that have a signed approved tabulated data sheet available may be used. The tabulated data sheets are prepared by the trench box manufacturer and are available through the Safety and Health Manager or the shop supervisor's office, and may be permanently attached to some boxes. The limits set forth by these data sheets must not be exceeded.
c. Trench boxes must be inspected by the immediate supervisor (Competent Person) prior to and during use, to guard against structural failures. Boxes requiring assembly must be done under the supervision of the designated Competent Person.
d. Trench boxes may be used in conjunction with sloping or benching in order to meet the required safe employee protection regulations.
e. Trench boxes may be undercut up to 2 ft. as long as overall tabulated data sheet limits are not exceeded and the soil the box is sitting on is stable.
f. Trench boxes must extend at least 18 in. above the top of slope of the excavation to prevent materials, earth, tools, etc. from falling into the work area. Generally, extra shields or plates may be used to provide the additional 18 in.

g. Personnel working in a trench box must always remain in the protected area(s) including during access and egress. Ladders, when used, must be inside the protected area.
h. A trench box must never be moved while personnel are in it or on it.

15.28 UNDERWATER WORK

15.28.1 General Policy

NOTE TO MANUAL PREPARER: The following sample section assumes a company nor regularly engaged in underwater work. If your company does work in these conditions, modify the following language accordingly.

a. All underwater work required will be done by the use of outside contractors licensed and qualified to perform such work in the state in which the project is located.

15.28.2 Use of Inflatable Pipe Plugs

NOTE TO MANUAL PREPARER: The following sample section assumes a utility contractor who may install storm and sanitary piping. If your company does not perform these operations, consider deleting or modifying the following language.

a. Although company policy in performing underwater work that requires personnel to be submerged to perform the necessary work is to use outside contractors, there may be occasions when the work requires use of an inflatable pipe plug where the plug is partially or fully submerged but to install and remove the plug, a person need not be submerged. In these cases, if the company chooses to perform the operation itself, the following procedures must be followed to safeguard our employees:
   i. Section 15.20.8 of the *Safety and Loss Control Program Manual,* "Life Jackets and Ring Buoys," will be implemented at the onset of operations and be maintained throughout.
   ii. Inflatable plug(s) are to be properly and prominently identified as to whether they are bypass or non-bypass type.
   iii. Non-bypass type plugs must be so identified with a label on or attached to the domehead.
   iv. Only persons trained in inflatable pipe plug use shall install/remove plug(s).

v. Plugs must be deflated by onshore methods.
vi. Equalize pressure in pipelines by pumping water into pipe through collection boxes; not through use of bypass piping.

15.29 WELDING AND TORCH OPERATIONS (HOT WORK)

15.29.1 General

a. No welding or torch operations are to be performed by any **(COMPANY)** subcontractor without a properly prepared hot work permit, in accordance with the requirements of Section 14.10 of the *Safety and Loss Control Program Manual.*
b. When possible, items to be welded, cut, heated, etc. shall be moved to a safe location free of combustible or flammable material. If this is not possible, then all combustibles/flammables that can be removed from the area will be moved and a positive means of confining arcs and sparks generated by the process will be taken and additional person(s) will be stationed as fire watch for the area(s) still exposed.
c. Appropriate fire-fighting equipment is to be available in close proximity of any welding and torch operations at all times.
d. Drums, tanks, and similar containers that have contained flammable or toxic material shall not be welded, cut, or heated until they have been made safe by water filling, thorough cleansing, or similar accepted practices. The container must also be ventilated during the welding, cutting, or heating process.
e. Proper eye protection must be worn during all welding and torch operations.
f. Proper ventilation is required for any welding or torch operations performed in a confined space.
g. Any welding or torch operations performed on metals of toxic compounds or coating such as zinc, stainless steel, lead, cadmium, chromium, and beryllium must be properly ventilated and/or proper respiratory protection must be worn by any person that could be exposed to fumes, vapors, and gasses created by the welding and torch cutting processes.

15.29.2 Arc Welding

a. Wherever it is practical, all arc welding operations shall be shielded to prevent direct light rays or sparks from contacting persons in the vicinity or from reaching areas normally used to travel through or into the vicinity. Where this is not practical, persons who must be in the area are to use proper eye and skin protection. Other persons who are not participating in the welding or torch operations are not to be allowed into the hazard zone.

b. Welders and other employees who are exposed to arc welding radiation must wear suitable clothing and protective apparel to prevent burns and other types of ultraviolet radiation damage to the skin.
c. Arc welding machines must be shut down when being moved or when they are not in continuous use.
d. Electrode holders left unattended must have electrodes removed and must not be left where they might contact employees or conducting objects.
e. Energized electrodes must never be dipped in water.
f. Welding hoods must be checked for light leaks periodically to ensure the user will be protected against eye flash burns.
g. Welders must guard against allowing materials adjacent to or behind them to reflect radiation back toward them or towards others in the area. Reflected radiation can cause skin burns and eye flash burns.

15.29.3 Gas Welding and Torch Operations

a. Compressed gas cylinders must always be secured from tipping or falling, whether in use, in storage or in transit. The cylinders must always be secured upright, except during times when actually being hoisted or carried.
b. Valve caps must be in place when cylinders are not in use. Valve caps must never be used for lifting the cylinder vertically.
c. Regulators must be removed when cylinders are not in use or are in transit, unless the cylinder is firmly secured on a special carrier designed for this purpose.
d. Cylinders being transported by a powered vehicle must be secured in an upright position.
e. Cylinders being hoisted must be secured on a pallet cradle or similar lifting platform.
f. Cylinders must be shielded or kept far enough away from welding or torch operations so that sparks, flames, or slag from the operation cannot reach them.
g. Cylinders containing oxygen or fuel gasses must not be taken into confined spaces.
h. Fuel gas cylinders requiring special wrenches shall have the wrench in place on the stem while the cylinder is in use, so that it is immediately available to close the cylinder valve.
i. Oxygen cylinders shall be stored a minimum of 20 ft. from fuel gas cylinders or must have an approved fire wall between them.
j. Oily or greasy substances must never be allowed to get near or come in contact

with the valve end of an oxygen cylinder or its related apparatus (hose, regulator). This includes hands, gloves, etc. that have oil or grease on them.

k. Torches must only be lit by approved strikers; never with matches, cigarette lighters, or hot work.

15.30 WIRE ROPES, CHAINS, ROPES, ETC.

a. Wire ropes, chains, ropes and other rigging equipment will be inspected prior to use and as necessary during use to assure their safety. Defective rigging equipment is to be removed from service immediately.

b. Job or shop hooks and links, or makeshift fasteners formed from bolts, rods, or other such attachments, are not to be used. When U-bolts are used for eye splices, the U-bolt will be applied so the "U" section is in contact with dead end of rope.

Sample Section 16

Equipment Safety and Maintenance

Section Use

The form and content of Sample Section 6 *Equipment Safety and Maintenance* is designed as an example of company instruction to its employees with respect to the material presented. It is intended to provide for each employee:

- A summary of general safety rules to be applied by company operations to the use of all equipment, and
- Checklists of preventative maintenance considerations to be followed by equipment operators to assist those operators in ensuring proper ongoing use of all equipment.

The sample section that follows is an example of a company safety and loss control program approach to the objectives indicated above. The sample language and forms are to be considered in the context of your own company and its manner of operation.

The safety and maintenance information presented is of general nature. The preparer of an actual Safety and Loss Control Program Manual may consider combining this sample section with the following Sample Section 17, and incorporating the safety and equipment checklists for each type of major equipment actually being used by your own company either directly in the company's Safety and Loss Control Program Manual, or in a separate *Equipment Safety & Maintenance Manual* that is coordinated with the complete Safety and Loss Control Program Manual.

Carefully consider these objectives during the review and editing process of the section material in order to customize the section content to the specific needs of your organization. Prepare final language that is consistent with decisions made by your own management in the determination of your own company program.

It is important to understand that safety rules and regulations change from time to time. Be sure to review the final draft of the section and the entire program manual with appropriate safety, insurance, and legal professionals prior to issue and use. Periodically review your program with company management and all appropriate professionals to be certain that the company program remains current with all requirements.

(Sample)

SECTION 16

Equipment Safety and Maintenance

COMPANY LOGO	Construction Safety and Loss Control Program Manual	Page: 16-2	Appr:
	16. EQUIPMENT SAFETY AND MAINTENANCE	Revision Date:	

16.1 EQUIPMENT SAFETY, PRE-START, AND MAINTENANCE

Equipment safety, reliability, and operating efficiency are no accident. Following simple rules of observation and reasonable care will reduce equipment down time, increase operating efficiency, and improve operator well being.

Every (**COMPANY**) equipment operator should be thoroughly familiar with the checklists of this section, and follow them each day that every piece of equipment is being operated.

The rules and checklists that follow are generic; they apply generally to all equipment. Accordingly, each equipment operator must also be thoroughly familiar with the specific safety and maintenance checklists provided for each specific piece of equipment used, and apply them daily—in addition to the rules and checklists of this section.

16.1.1 General Safety Rules

1. No persons other than (**COMPANY**) personnel are allowed to operate any (**COMPANY**) equipment under any circumstances.
2. Only qualified operators are allowed to operate the equipment.
3. Operators are to be familiar with the location and purpose of all controls, instruments, indicator lights, and labels.
4. A first aid kit must be fastened to all major equipment.
5. A fully charged fire extinguisher must be mounted conveniently. Operators must be familiar with proper use of all fire-fighting equipment.
6. Fairly tight, properly fitting clothing and safety equipment should be worn by operators. Loose clothing should be avoided.
7. High-pressure fluids that can penetrate skin and cause injury should be avoided.
   a. Relieve pressure before disconnecting hydraulic or other lines.
   b. Tighten all connections before applying pressure.
   c. Keep hands and body away from pinholes and nozzles that eject fluids under high pressure.
   d. Use a piece of cardboard to search for leaks. Do not use your hand.
   e. ***If ANY fluid is ejected into the skin, it must be surgically removed within a few hours by a physician, or gangrene may result with this type of injury.***
8. Suitable hearing protective device such as earmuffs or earplugs should be worn in order to protect against loud noise.

16.1.2 Operation Safety Rules

1. Use handrails and steps to enter and leave the operator's station. Do not use the steering wheel.
2. Keep handrails, steps, floor, and controls free of water, grease, and dirt.
3. Do not operate any equipment in an unsafe condition. Put a tag on the steering wheel or other appropriate high-visibility location to warn others of any unsafe condition.
4. Before starting or operating any equipment:
   a. Check the condition of the equipment. Refer to Pre-Start Inspection Checklist of Section 16.1.3 below.
   b. Be sure there is sufficient ventilation.
   c. Know the correct starting and stopping procedure.
   d. Sit in the operator's seat.
   e. Clear the work area of people and obstacles.
   f. Check the service brakes and parking brake.
5. Be sure engine is running and foot brakes are operating before releasing the parking brake.
6. Do not allow riders on the equipment.
7. Drive slowly in congested areas, over rough ground, and on slopes and curves.
8. Do not drive near the edges of ditches and excavations.
9. Keep loading areas smooth and flat.
10. Check locations of utilities, cables, gas lines, water mains, etc. before digging.
11. Keep away from power lines at all costs. Do not touch power lines with any part of the equipment.
12. Carry buckets and loads as low as possible for better stability and visibility.
13. Keep equipment in gear when going down steep grades.
14. Use accessory lights and devices to warn operators of other vehicles.
15. Position backhoe booms on uphill side when driving across hillsides.
16. Set stabilizers before operating any backhoe equipment.
17. Use care when raising stabilizers; they may be the only restraint preventing the equipment from rolling into an excavation.
18. Do not dig under stabilizers.
19. Avoid swinging any backhoe bucket in the downhill direction. Always dump on uphill side, if possible.
20. Before dismounting the equipment:
    a. Engage parking brake.
    b. Lower all equipment to the ground.
    c. Stop engine.

d. Release hydraulic pressure; turn steering wheel back and forth, move hydraulic control levers until the equipment does not move.

16.1.3 Pre-Start Inspection Checklist

Follow the checklist below to inspect all equipment before your start it—***each day that the equipment is used***.

1. Vandalism. Check to see that:
   a. Smokestacks and exhaust pipes are clear of debris and obstructions.
   b. Fuel, water, gas, and oil filters have not been tampered with.
   c. Lights and glass are not broken or loosened.
   d. Gauges are not damaged.
   e. Wires have not been cut.
   f. The equipment looks in good overall condition.
2. Tires and Wheels
   a. Inspect for loose or missing bolts.
   b. Check tire pressure.
3. Operator's Station
   a. Be sure it's clean.
   b. Check pedals for freedom of movement.
4. Hydraulic System. Check:
   a. Oil levels
   b. For leaks, kinked lines, and lines or hoses that rub against each other or other parts.
5. Engine Compartment. Check:
   a. Engine oil level.
   b. Transmission oil level.
   c. Fuel filter for sediments.
   d. Air cleaner.
   e. Radiator coolant level and clean radiator.
6. Lubrication. Check:
   a. Lubrication points shown in the respective equipment service manual.
7. Electrical System. Check:
   a. For worn or frayed wires and loose connections.
8. Protective Devices. Check:
   a. Guards, canopy, shields, seat belt.
9. Booms, buckets, structural components. Check:
   a. Bent, broken, or missing parts.

16.1.4 Service Safety Rules

1. Put a support under all raised equipment.
2. Before beginning service:
   a. Review requirements for hard hat, safety shoes, safety glasses or goggles, gloves, reflective vest, ear protectors, and respirator.
   b. Be sure that service is approved by the company equipment manager.
   c. Understand all procedures.
   d. Stop all equipment.
   e. Stop the engine (unless necessary for service).
3. Before welding anywhere on equipment or working on the engine or electrical systems, disconnect the battery ground wires.
4. Before working on the hydraulic system:
   a. Release all pressure.
   b. Loosen fittings slowly.
5. Do not smoke:
   a. When you fill the fuel tank.
   b. When you work on the fuel system.
   c. When you handle fuels or lubricants.
6. Do not fill the fuel tank when the engine is running.
7. Guard against eye injury when hammering.
8. Do not lubricate or work on equipment when it is in motion.
9. Avoid high pressure fluids:
   a. Relieve pressure before disconnecting hydraulic or other lines.
   b. Tighten all connections before applying pressure.
   c. Keep hands and body away from pinholes and nozzles that eject fluids under high pressure.
   d. Use a piece of cardboard to search for leaks—*do not use your hand.*
10. Engine coolant:
   a. Only add coolant to the radiator when the engine is stopped or running at slow idle.
   b. Do not remove cap unless engine is cool.
   c. Release all pressure before removing cap, and loosen slowly.
11. Tires:
   a. Do not attempt to mount tires without the proper equipment and experience to perform the job safely. ***Failure to follow proper procedure can produce an explosion which may result in serious bodily injury or death.***
   b. Be sure all tire rims are correctly assembled and interlocking before inflating tires.
   c. Use an inflation cage, safety cables, or some such protective device during inflation.

COMPANY LOGO	Construction Safety and Loss Control Program Manual	Page: 16-6	Appr:
	16. EQUIPMENT SAFETY AND MAINTENANCE	Revision Date:	

16.1.5 Fire Prevention Maintenance

1. Daily Pre-Start Maintenance:
   a. Check fire extinguisher for correct charge.
   b. Open all access hoods and shields. Remove all trash from all areas inside these compartments from:
      - Exhaust manifold, turbocharger, and muffler.
      - Bottom guards and under engine.
      - Sides of engine.
      - Radiator and oil cooler.
      - Batteries.
      - Hydraulic lines.
      - Fuel tank.

   c. Check for leaking fuel lines, hydraulic lines or fittings. Tighten loose fittings. Replace bent or kinked lines.
   d. Clean trash from grilles.
   e. Clean trash from cab areas.
   f. Be sure all doors and grilles are in place.
2. Shut-Down:
   a. Temperature in engine compartment may go up immediately after stopping engine. Be on guard for fires; especially when refueling.
   b. Wait until the engine has cooled before filling the fuel tank.
   c. Do not smoke when refueling.

Sample Section 17

(MANUFACTURER AND MODEL NUMBERS) —SAMPLE EQUIPMENT SECTION

Section Use

The form and content of Section 17 *(Manufacturer and Model Numbers)—Sample Equipment Section* is designed as an example of company instruction to its employees with respect to the material presented. It is intended to provide for each employee:

- Inspection, maintenance, and operating procedures and checklists for each piece of equipment owned and operated by the company

The sample section that follows is an example of a company safety and loss control program approach to the objectives indicated above. The sample language and forms are to be considered in the context of your own company and its manner of operation.

The safety and maintenance information presented is of general nature. The preparer of an actual Safety and Loss Control Program Manual should consider securing similar information for equipment owned and/or operated. Also consider combining this sample section with the previous sample Section 16, and incorporating the safety and equipment checklists for each type of major equipment actually being used by your own company either directly in the company's final Safety and Loss Control Program Manual, or in a separate *Equipment Safety and Maintenance Manual* that is coordinated with the complete Safety and Loss Control Program Manual.

Carefully consider these objectives during the review and editing process of the section material in order to customize the section content to the specific needs of your organization. Prepare final language that is consistent with decisions made by your own management in the determination of your own company program.

It is important to understand that safety rules and regulations change from time to time. Be sure to review the final draft of the section and the entire program manual with appropriate safety, insurance, and legal professionals prior to issue and use. Periodically review your program with company management and all appropriate professionals to be certain that the company program remains current with all requirements.

(Sample)

SECTION 17

(Manufacturer & Model Numbers)—Sample Equipment Section

17.1 EQUIPMENT SAFETY, PRE-START, AND MAINTENANCE

NOTE TO MANUAL PREPARER: The entire sample section that follows is an example of information and checklists provided for one type of equipment. The format should be repeated for each individual piece of equipment used regularly by the company. The information should then be modified as required and supplemented as necessary by each respective equipment manufacturer.

Equipment safety, reliability, and operating efficiency are no accident. Following simple rules of observation and reasonable care will reduce equipment down time, increase operating efficiency, and improve operator well being.

Every (**COMPANY**) equipment operator should be thoroughly familiar with the checklists of this section, and follow them each day that every piece of equipment is being operated.

In addition, each equipment operator must also be thoroughly familiar with the generic safety and maintenance checklists provided in Section 16 of this Manual, and apply them daily—*in addition* to the rules and checklists of this section for specific equipment.

17.2 (MANUFACTURER AND MODEL NUMBERS) BACKHOES—CHECKLISTS AND INSPECTIONS

17.2.1 Daily Inspection

1. Walk around machine and check for leaks before operating.
2. Write down hour meter reading, and put on time sheet.
3. Check engine oil.
4. Check water in radiator.
5. Check hydraulic oil.
6. Clear dirt out of dust collectors.
7. Check fuel levels and drain from fuel tank.
8. Clean dirt out of cab and from around control pedals.

17.2.2 Before Operating Machine

1. Start engines and let warm up for (15) minutes while greasing bucket and stick points.

2. Grease bucket pins, twice daily: Once in the morning; once after lunch.
3. Grease stick and boom pivot points once per day.
4. Check travel alarm. If it does not work, advise the foreman or call the shop immediately.

17.2.3 Weekly (Friday) Inspection

1. Check battery water levels, clean top of batteries..
2. Grease swing bearing.
3. Check hydraulic pump gear box. If oil is needed, contact the shop immediately.
4. Check swing drive oil level on 245 Cat.

17.2.4 Shut Down Procedure

1. Allow engine to cool down at an idle for (10) minutes.
2. Park machine in a safe area; lock all doors and panels.
3. Relieve all hydraulic pressure. Lower all booms and arm to the ground.

17.2.5 Do Not's

1. ***DO NOT*** level dirt with backhoe bucket by swinging the bucket from side to side, knocking the dirt down.
2. ***DO NOT*** travel any distance with the travel motors (sprockets) in front (3–1/2 times track wear in reverse).
3. ***DO NOT*** operate the swing bearing or center joint in water or mud. If this accidentally happens, contact the shop immediately.
4. ***DO NOT*** operate near power lines.

17.3 OPERATOR ACKNOWLEDGMENT FORM

The operator acknowledgment on the following page reiterates the checklists of this section, and confirms the operator's awareness of expected company procedure.

Before being allowed to operate any **(COMPANY)** equipment, every operator must:

1. Read and thoroughly understand Section 16, ***Equipment Safety and Maintenance Manual.***

COMPANY LOGO	Construction Safety and Loss Control Program Manual	Page: 17-4	Appr:
	17. SAMPLE EQUIPMENT SECTION	Revision Date:	

2. Read and thoroughly understand this section, including the complete equipment manual included herein.
3. Sign the Operator Acknowledgment, confirming the operator's understanding and agreement to follow all established safety, operation, and maintenance procedures.

Operator Acknowledgment
(Manufacturer & Model Numbers)—Checklists and Inspections

Daily Inspection

1. Walk around machine and check for leaks before operating.
2. Write down hour meter reading, and put on time sheet.
3. Check engine oil.
4. Check water in radiator.
5. Check hydraulic oil.
6. Clear dirt out of dust collectors.
7. Check fuel levels and drain from fuel tan.
8. Clean dirt out of cab and from around control pedals.

Before Operating Machine

1. Start engines and let warm up for (15) minutes while greasing bucket and stick points.
2. Grease bucket pins, twice daily: Once in the morning; once after lunch.
3. Grease stick and boom pivot points once per day.
4. Check travel alarm. If it does not work, advise the foreman or call the shop immediately.

Weekly (Friday) Inspection

1. Check battery water levels, clean top of batteries.
2. Grease swing bearing.
3. Check hydraulic pump gear box. If oil is needed, contact the shop immediately.
4. Check swing drive oil level.

Shut Down Procedure

1. Allow engine to cool down at an idle for (10) minutes.
2. Park machine in a safe area; lock all doors and panels.

Do Not's

1. ***DO NOT*** level dirt with backhoe bucket by swinging the bucket from side to side, knocking the dirt down.

2. ***DO NOT*** travel any distance with the travel motors (sprockets) in from (3–1/2 times track wear in reverse).
3. ***DO NOT*** operate the swing bearing or center joint in water or mud. If this accidentally happens, contact the shop immediately.
4. ***DO NOT*** operate near power lines.

I understand all of the check points described within this list, have read and understand Section 16 *Equipment Safety and Maintenance* of the **(COMPANY)** *Safety and Loss Control Program Manual,* and have asked all pertinent questions relating to each piece of equipment operated by me. I have read and understand all applicable safety rules and guidelines, and understand that not following these rules and guidelines may be ground for dismissal.

Operator: Signature: ______________________________

Name (Print): ______________________________

Employee #: ______________ Date: __________

17.4 EQUIPMENT MANUAL

NOTE TO MANUAL PREPARER: Include either a direct reference to the location of the complete equipment manual for the specific equipment, or consider including a copy of the complete equipment manual itself directly in this area of the sample section, if the manual is not voluminous.

Appendix

A-1 GENERAL DESCRIPTION

This Appendix is provided as a supplement to the complete sample *Safety and Loss Control Program Manual* as it has been presented. It is a compilation of selected training requirements and guidelines that have been prepared by the U.S. Department of Labor Occupational Safety and Heath Administration specifically to assist companies in the distillation process of converting stated compliance requirements of the OSHA standard into practical operating activities.

The U.S. Department of Labor has provided these types of material guidelines for many industries, and in a manner that addresses working conditions throughout many regional environments. In an effort to improve the efficiency of the use of this type of information for the members of the construction industry, this Appendix is the result of isolating and consolidating those requirements and guidelines that directly address our own construction industry.

For anyone charged with the responsibility to prepare, maintain, present, and communicate the objectives of this sample *Construction Safety and Loss Control Program Manual*, it will prove to be a great advantage to review this Appendix prior to and throughout the preparation of any kind of actual programs and program materials for your own company use, in order to preserve the ideals of the OSHA standard, and to ensure that every portion of the actual final company program remains consistent with the highest possible standards of performance.

The following is derived directly from materials developed b the U.S. Department of Labor Occupational Safety and Health Administration. It is not to be considered as any substitute for any provisions of the Occupational Safety and Health Act of 1970, or for any standards issued by OSHA, but is to be considered as a concise informational guide.

A-2 INTRODUCTION

Many standards promulgated by the Occupational Safety and Health Administration (OSHA) explicitly require the employer to train employees in the safety and health aspects of their jobs. Other OSHA standards make it the employer's responsibility to limit certain job assignments to employees who are certified, or qualified—meaning that they have had special previous training, in or out of the workplace. The term "designated personnel" means selected or assigned by the employer or the employee's representative as being qualified to perform specific duties. These requirements reflect OSHA's belief that training is an essential part of every employer's safety and health program for protecting workers from injuries and illnesses. Many researchers conclude that those who are new on the job have a higher rate of accidents and injuries than more experienced workers. If ignorance of specific job hazards and of proper work practices is even partly to blame for this higher injury rate, then training will help to provide a solution.

As an example of the trend in OSHA safety and health training requirements, the Process Safety Management of Highly Hazardous Chemicals Standard (Title 29 Code of Federal Regulations Part 1910, 119) contains several training requirements. This standard was promulgated under the requirements of the Clean Air Act Amendments of 1990. The Process Safety Management Standard requires the employer to evaluate or verify that employees comprehend the training given to them. This means that the training to be given must have established goals and objectives regarding what is to be accomplished. Subsequent to the training, an evaluation would be conducted to verify that the employees understood the subjects presented or acquired the desired skills. If the established goals and objectives of the training program were not achieved as expected, the employer then would revise the training program to make it more effective, or conduct more frequent refresher training or some combination of these. These requirements of the Process Safety Management Standard follow the concepts embodied in the OSHA training guidelines contained in this booklet.

The length and complexity of OSHA standards may make it difficult to find all the references to training. So, to help employers, safety and health professionals, training directors, and others with a need to know, OSHA's training-related requirements have been excerpted and collected in this booklet. Requirements for posting information, warning signs, labels, and the like are excluded, as are most references to the qualifications of people assigned to test workplace conditions or equipment.

It is usually a good idea for the employer to keep a record of all safety and health training. Records can provide evidence of the employer's good faith and compliance with OSHA standards. Documentation can also supply an answer to one of the first questions an accident investigator will ask: Was this injured employee trained to do the job?

Training in the proper performance of a job is time and money well spent, and the employer might regard it as an investment rather than an expense. An effective program of safety and health training for workers can result in fewer injuries and illnesses, better morale, and lower insurance premiums, among other benefits.

Readers with questions concerning worker safety and health training should contact their OSHA Regional Office.

A-3 VOLUNTARY TRAINING GUIDELINES

A-3.1 Introduction

The Occupational Safety and Health Act of 1970 does not address specifically the responsibility of employers to provide health and safety information and instruction to employees, although Section 5(a)(2) does require that each employer shall comply with occupational safety and health standards promulgated under this Act. However, more than 100 of the Act's current standards do contain training requirements.

Therefore, the Occupational Safety and Health Administration has developed voluntary training guidelines to assist employers in providing the safety and health information and instruction needed for their employees to work at minimal risk to themselves, to fellow employees, and to the public.

The guidelines are designed to help employers to: 1) determine whether a workaday problem can be solved by training; 2) determine what training, if any, is needed; 3) identify goals and objectives for the training; 4) design learning activities; 5) conduct training; 6) determine the effectiveness of the training; and 7) revise the training program based on feedback from employees, supervisors, and others.

The development of the guidelines is part of an agency-wide objective to encourage cooperative, voluntary safety and health activities among OSHA, the business community, and workers. These voluntary programs include training and education, consultation, voluntary protection programs, and abatement assistance.

A-3.2 Training Model

The guidelines provide employers with a model for designing, conducting, evaluating, and revising training programs. The training model can be used to develop training programs for a variety of occupational safety and health hazards identified at the workplace. Additionally, it can assist employers in their efforts to meet the training requirements in current or future occupational safety and health standards.

A training program designed in accordance with these guidelines can be used to supplement and enhance the employer's other education and training activities. The guidelines afford employers significant flexibility in the selection of content and training program design. OSHA encourages a personalized approach to the informational and instructional programs at individual worksites, thereby enabling employers to provide the training that is most needed and applicable to local working conditions.

Assistance with training programs or the identification of resources for training is available through such organizations as OSHA full service Area Offices, state agencies which have their own OSHA approved occupational safety and health programs, OSHA funded state on-site consultation programs for employers, local safety councils, the OSHA Office of Training and Education, and OSHA funded New Directions grantees.

A-3.3 Review Commission Implications

OSHA does not intend to make the guidelines mandatory. They should not be used by employers as a total or complete guide in training and education matters which can result in enforcement proceedings before the Occupational Safety and Health Review Commission in cases which involve alleged violations of training requirements contained in OSHA standards.

The adequacy of employee training may also become an issue in contested cases where the affirmative defense of unpreventable employee misconduct is raised. Under case law well established in the Commission and the courts, an employer may successfully defend against an otherwise valid citation by demonstrating that all feasible steps were taken to avoid the occurrence of the hazard, and that actions of the employee involved in the violation were a departure from a uniformly and effectively enforced work rule of which the employee had either actual or constructive knowledge.

In either type of case, the adequacy of the training given to employees in connection with a specific hazard is a factual matter which can be decided only by considering all the facts and circumstances surrounding the alleged violation. The general guidelines in this publication are not intended, and cannot be used, as evidence of the appropriate level of training in litigation involving either the training requirements of OSHA standards or affirmative defenses based upon employer training programs.

A-3.4 Training Guidelines

OSHA's training guidelines follow a model that consists of:

A. Determining if Training if Needed
B. Identifying Training Needs
C. Identifying Goals and Objectives

D. Developing Learning Activities
E. Conducting the Training
F. Evaluating Program Effectiveness
G. Improving the Program
H. Matching Training to Employees
I. Identifying Employers at Risk
J. Training Employees at Risk

The model is designed to be one that even the owner of a business with very few employees can use without having to hire a professional trainer or purchase expensive training materials. Using this model, employers or supervisors can develop and administer safety and health training programs that address problems specific to their own business, fulfill the learning needs of their own employees, and strengthen the overall safety and health program of the workplace.

A-3.5 Determining If Training is Needed

The first step in the training process is a basic one: to determine whether a problem can be solved by training. Whenever employees are not performing their jobs properly, it is often assumed that training will bring them up to standard. However, it is possible that other actions (such as hazard abatement or the implementation of engineering controls) would enable employees to perform their jobs properly.

Ideally, safety and health training should be provided before problems or accidents occur. This training would cover both general safety and health rules and work procedures, and would be repeated if an accident or near-miss incident occurred.

Problems that can be addressed effectively by training include those that arise from lack of knowledge of a work process, unfamiliarity with equipment, or incorrect execution of a task. Training is less effective (but still can be used) for problems arising from an employee's lack of motivation or lack of attention to the job. Whatever its purpose, training is most effective when designed in relation to the goals of the employer's total safety and health program.

A-3.6 Identifying Training Needs

If the problem is one that can be solved, in whole or in part, by training, then the next step is to determine what training is needed. For this, it is necessary to identify what the employee is expected to do and in what ways, if any, the employee's performance is deficient. This information can be obtained by conducting a job analysis which pinpoints what an employee needs to know in order to perform a job.

When designing a **new training** program, or preparing to instruct an employee in an unfamiliar procedure or system, a job analysis can be developed by examining engineering data on new equipment or the safety data sheets on unfamiliar substances. The content of the specific federal or state OSHA standards applicable to a business can also provide direction in developing training content. Another option is to conduct a Job Hazard Analysis (see OSHA 3071, same title, 1981). This is a procedure for studying and recording each step of a job, identifying existing or potential hazards, and determining the best way to perform the job in order to reduce or eliminate the risks. Information obtained from a Job Hazard Analysis can be used as the content for the training activity.

If an employer's learning needs can be met by revising an *existing* training program rather than developing a new one, or if the employer already has some knowledge of the process or system to be used, appropriate training content can be developed through such means as:

1. Using company accident and injury records to identify how accidents occur and what can be done to prevent them from recurring.
2. Requesting employees to provide, in writing and in their own words, descriptions of their jobs. These should include the tasks performed and the tools, materials and equipment used.
3. Observing employees at the workaday level as they perform tasks, asking about the work, and recording their answers.
4. Examining similar training programs offered by other companies in the same industry, or obtaining suggestions from such organizations as the National Safety Council (which can provide information of Job Hazard Analysis), the Bureau of Labor Statistics, OSHA-approved state programs, OSHA full-service Area Offices, OSHA-funded state consultation programs, or the OSHA Office of Training and Education.

The employees themselves can provide valuable information on the training they need. Safety and health hazards can be identified through the employees' responses to such questions as whether anything about their jobs frightens them, if they have had any near-miss incidents, if they feel they are taking risks, or if they believe that their jobs involve hazardous operations or substances.

Once the kind of training that is needed has been determined, it is equally important to determine what kind of training is **not** needed. Employees should be made aware of all the steps involved in a task or procedure, but training should focus on those steps on which improved performance is needed. This avoids unnecessary training and tailors the training to meet the needs of the employees.

A-3.7 Identifying Goals and Objectives

Once the employees' training needs have been identified, employers can then prepare objectives for the training. Instructional objectives, if clearly stated, will tell employers what they want their employees to do, to do better, or to stop doing.

Learning objectives do not necessarily have to be written, but in order for the training to be as successful as possible, clear and measurable objectives should be thought out before the training begins. For an objective to be effective it should identify as precisely as possible what the individuals will do to demonstrate that they have learned, or that the objective has been reached. They should also describe the important conditions under which the individual will demonstrate competence and define what constitutes acceptable performance.

Using specific, action-oriented language, the instructional objectives should describe the preferred practice or skill and its observable behavior. For example, rather than using the statement: "The employee will understand how to use a respirator" as an instructional objective, it would be better to say: "The employee will be able to describe how a respirator works and when it should be used." Objectives are most effective when worded in sufficient detail that other qualified persons can recognize when the desired behavior is exhibited.

A-3.8 Developing Learning Activities

Once employers have stated precisely what the objectives for the training program are, then learning activities can be identified and described. Learning activities enable employees to demonstrate that they have acquired the desired skills and knowledge. To ensure that employees transfer the skills or knowledge from the learning activity to the job, the learning situation should simulate the actual job as closely as possible. Thus, employers may want to arrange the objectives and activities in a sequence which corresponds to the order in which the tasks are to be performed on the job, if a specific process is to be learned. For instance, if an employee must learn the beginning processes of using a machine, the sequence might be: (1) to check that the power source is connected; (2) to ensure that the safety devices are in place and are operative; (3) to know when and how to throw the switch; and so on.

A few factors will help to determine the type of learning activity to be incorporated into the training. One aspect is the training resources available to the employer. Can a group training program that uses an outside trainer and firm be organized, or should the employer personally train the employees on a one-to-one basis? Another factor is the kind of skills or knowledge to be learned. Is the learning oriented toward physical skills (such as the use of special tools) or toward mental processes and attitudes? Such factors will influence the type of learning activity designed by employers. The training activity can be group-oriented, with lectures, role play, and demonstrations, or designed for the individual as with self-paced instruction.

The determination of methods and materials for the learning activity can be as varied as the employer's imagination and available resources will allow. The employer may want to use charts, diagrams, manuals, slides, films, viewgraphs (overhead transparencies), videotapes, audiotapes, or simply blackboard and chalk, or any combination of these and other instructional aids. Whatever the method of instruction, the learning activities should be developed in such a way that the employees can clearly demonstrate that they have acquired the desired skills or knowledge.

A-3.9 Conducting the Training

With the completion of the steps outlined above, the employer is ready to begin conducting the training. To the extent possible, the training should be presented so that its organization and meaning are clear to the employees. To do so, employers or supervisors should: **(1)** provide overviews of the material to be learned; **(2)** relate, wherever possible, the new information or skills to the employee's goals, interests, or experience; and **(3)** reinforce what the employees learned by summarizing the program's objectives and the key points of information covered. These steps will assist employers in presenting the training in a clear, unambiguous manner.

In addition to organizing the content, employers must also develop the structure and format of the training. The content developed for the program, the nature of the workplace or other training site, and the resources available for training will help employers determine for themselves the frequency of training activities, the length of the sessions, the instructional techniques, and the individual(s) best qualified to present the information.

In order to be motivated to pay attention and learn the material that the employer or supervisor is presenting, employees must be convinced of the importance and relevance of the material. Among the ways of developing motivation are: **(1)** explaining the goals and objectives of instruction; **(2)** relating the training to the interests, skills, and experiences of the employees; **(3)** outlining the main points to be presented during the training session(s); and **(4)** pointing out the benefits of training (e.g., the employee will be better informed, more skilled, and thus more valuable both on the job and on the labor market; or the employee will, if he or she applies the skills and knowledge learned, be able to work at reduced risk).

An effective training program allows employees to participate in the training process and to practice their skills or knowledge. This will help to ensure that they are learning the required knowledge or skills and permit correction if necessary.

Employees can become involved in the training process by participating in discussions, asking questions, contributing their knowledge and expertise, learning through hands-on experiences, and through role-playing exercises.

A-3.10 Evaluating Program Effectiveness

To make sure that the training program is accomplishing its goals, an evaluation of the training can be valuable. Training should have, as one of its critical components, a method of measuring the effectiveness of the training. A plan for evaluation the training session(s), either written or thought out by the employer, should be developed when the course objectives and content are developed. It should not be delayed until the training has been completed. Evaluation will help employers or supervisors determine the amount of learning achieved and whether an employee's performance has improved on the job. Among the methods of evaluating training are: **(1) Student opinion.** Questionnaires or informal discussions with employees can help employers determine the relevance and appropriateness of the training program; **(2) Supervisors' observation.** Supervisors are in good positions to observe an employee's performance both before and after the training and note improvements or changes; and **(3) Workplace improvements.** The ultimate success of a training program may be changes throughout the workplace that result in reduced injury or accident rates.

However it is conducted, an evaluation of training can give employers the information necessary to decide whether or not the employees achieved the desired results, and whether the training session should be offered again at some future date.

A-3.11 Improving the Program

If, after evaluation, it is clear that the training did not give the employees the level of knowledge and skill that was expected, then it may be necessary to revise the training program or provide periodic retraining. At this point, asking questions of employees and of those who conducted the training may be of some help. Among the questions that could be asked are: **(1)** Were parts of the content already known and, therefore, unnecessary? **(2)** What material was confusing or distracting? **(3)** Was anything missing from the program? **(4)** What did the employees learn, and what did they fail to learn?

It may be necessary to repeat steps in the training process, that is, to return to the first steps and retrace one's way through the training process. As the program is evaluated, the employer should ask: **(1)** If a job analysis was conducted, was it accurate? **(2)** Was any critical feature of the job overlooked? **(3)** Were the important gaps in knowledge and skill included? **(4)** Was material already known by the employees intentionally omitted? **(5)** Were the instructional objectives presented clearly and concretely? **(6)** Did the objectives state the level of acceptable performance that was expected of employees? **(7)** Did the learning activity simulate the actual job? **(8)** Was the learning activity appropriate for the kinds of knowledge and skills required on the job? **(9)** When the training was presented, was the organization of the material and its meaning made clear? **(10)** Were the employees motivated to learn? **(11)** Were the employees allowed to participate actively in the training process? **(12)** Was the employer's evaluation of the program thorough?

A critical examination of the steps in the training process will help employers to determine where course revision is necessary.

A-3.12 Matching Training to Employees

While all employees are entitled to know as much as possible about the safety and health hazards to which they are exposed, and employers should attempt to provide all relevant information and instruction to all employees, the resources for such an effort frequently are not, or are not believed to be, available. Thus, employers are often faced with the problem of deciding who is in the greatest need of information and instruction.

One way to differentiate between employees who have priority needs for training and those who do not is to identify employee populations which are at higher levels of risk. The nature of the work will provide an indication that such groups should receive priority for information on occupational safety and health risks.

A-3.13 Identifying Employees at Risk

One method of identifying employee populations at high levels of occupational risk (and thus in greater need of safety and health training) is to pinpoint hazardous occupations. Even within industries which are hazardous in general, there are some employees who operate at greater risk than others. In other cases the hazardousness of an occupation is influenced by the conditions under which it is performed, such as noise, heat or cold, or safety or health hazards in the surrounding area. In these situations, employees should be trained not only on how to perform their job safely but also on how to operate within a hazardous environment.

A second method of identifying employee populations at high levels of risk is to examine the incidence of accidents and injuries, both within the company and within the industry. If employees in certain occupational categories are experiencing higher accident and injury rates than other employees, training may be one way to reduce that rate. In addition, thorough accident investigation can identify not only specific employees who could benefit from training but also identify company-wide training needs.

Research has identified the following variables as being related to a disproportionate share of injuries and illnesses at the workaday on the part of employees:

1. The age of the employee (younger employees have higher incidence rates).
2. The length of time on the job (new employees have higher incidence rates).
3. The size of the firm (in general terms, medium size firms have higher incidence rates than smaller or larger firms).
4. The type of work performed (incidence and severity rates vary significantly by SIC Code).
5. The use of hazardous substances (by SIC Code).

These variables should be considered when identifying employee groups for training in occupational safety and health.

In summary, information is readily available to help employers identify which employees should receive safety and health information, education and training, and who should receive it before others. Employers can request assistance in obtaining information by contacting such organizations as OSHA area offices, the Bureau of Labor Statistics, OSHA approved state programs, state on-site consultation programs, the OSHA Office of Training and Education, or local safety councils.

A-3.14 Training Employees at Risk

Determining the content of training for employee populations at higher levels of risk is similar to determining what any employee needs to know, but more emphasis is placed on the requirements of the job and the possibility of injury. One useful tool for determining training content from job requirements is the Job Hazard Analysis described earlier. This procedure examines each step of a job, identifies existing or potential hazards, and determines the best way to perform the job in order to reduce or eliminate the hazards. Its key elements are: **(1)** job description; **(2)** job location; **(3)** key steps (preferably in the order in which they are performed); **(4)** tools, machines and materials used; **(5)** actual and potential safety and health hazards associated with these key job steps; and **(6)** safe and healthful practices, apparel, and equipment required for each job step.

Material Safety Date Sheets (MSDSs) can also provide information for training employees in the safe use of materials. These data sheets, developed by chemical manufacturers and importers, are supplied with manufacturing or construction materials and describe the ingredients of a product, its hazards, protective equipment to be used, safe handling procedures, and emergency first aid responses. The information contained in these sheets can help employers identify employees in need of training (i.e., workers handling substances described in the sheets) and train employees in safe use of the substances. Material Safety Data Sheets are generally available from suppliers, manufacturers of the substances, employers or trade associations. MSDSs are particularly useful for those employers who are developing training on chemical use as required by OSHA's Hazard Communication Standard.

A-3.15 Conclusion

In an attempt to assist employers with their occupational health and safety training activities, OSHA has developed a set of training guidelines in the form of a model. This model is designed to help employers develop instructional programs as part of their total education and training effort. The model addresses the questions of who should be trained, on what topics, and for what purposes. It also helps employers determine how effective the program has been and enables them to identify employees who are in greatest need of education and training. The model is general enough to be used in any area of occupational safety and health training, and allows employers to determine for themselves the content and format of training. Use of this model in training activities is just one of may ways that employers can comply with the OSHA standards that relate to training and enhance the safety and health of other employees.

A-4 GENERAL INDUSTRY TRAINING REQUIREMENTS

A-4.1 General

The following training requirements have been excerpted from Title 29, Code of Federal Regulations Part 1910. Note that additional training requirements may appear in certain other standards (ANSI, NFPS, etc.) adopted by reference in Part 1910 and therefore mandatory.

Employee Emergency Plans and Fire Prevention Plans
1910.38(a)(5)(i), (ii), and (iii)

(i) Before implementing the emergency action plan, the employer shall designate and train a sufficient number of persons to assist in the safe and orderly emergency evacuation of employees.

(ii) The employer shall review the plan with each employee covered by the plan at the following times:

(a) Initially when the plan is developed

(b) Whenever the employee's responsibilities or designated actions under the plan change, and

(c) Whenever the plan is changed.

(iii) The employer shall review with each employee upon initial assignment those parts of the plan which the employee must know to protect the employee in the event of an emergency. The written plan shall be kept at the workplace and made available for employee review. For those employers with 10 or fewer employees the plan may be communicated orally to employees and the employer need not maintain a written plan.

1910.(b)(4)(i) and (ii)

(i) The employer shall apprise employees of the fire hazards of the materials and processes to which they are exposed.

(ii) The employer shall review with each employee upon initial assignment those parts of the fire prevention plan which the employee must know to protect the employee in the event of an emergency. The written plan shall be kept in the workplace and made available for employee review. For those employers with 10 or fewer employees, the plan may be communicated orally to employees and the employer need not maintain a written plan.

Operations-Training 1910.66(i) (ii through v)

(i) Working platforms shall be operated only by persons who are proficient in the operation, safe use and inspection of the particular working platform to be operated.

(ii) All employees who operate working platforms shall be trained in the following:

(A) Recognition of, and preventive measures for, the safety hazards associated with their individual work tasks.

(B) General recognition and prevention of safety hazards associated with the use of working platforms, including the provisions in the section relating to the particular working platform to be operated.

(C) Emergency action plan procedures required in paragraph **(e)(9)** of this section.

(D) Work procedures required in paragraph **(i)(1)(iv)** of this section.

(E) Personal fall arrest system inspection, care, use and system performance.

(iii) Training of employees in the operation and inspection of working platforms shall be done by a competent person.

(iv) Written procedures for the operation, safe use and inspection of working platforms shall be provided for employee training. Pictorial methods of instruction may be used in lieu of written work procedures, if employee communication is improved using this method. The operating manuals supplied by manufacturers for platform system components can serve as the basis for these procedures.

(v) The employer shall certify that employees have been trained in operating and inspecting a working platform by preparing a certification record which includes the identity of the person trained, the signature of the person trained, the signature of the employer or the person who conducted the training and the date that training was completed. The certification record shall be prepared at the completion of the training required in paragraph (i)(1)(ii) of this section, and shall be maintained in a file for the duration of the employee's employment. The certification record shall be kept readily available for review by the Assistant Secretary of Labor or the Assistant Secretary's representative.

Care and Use Appendix C, Sec. 1 1910.66(e)(v)(9)

(9) Before using a personal fall arrest system, and after any component or system is changed, employees shall be trained in accordance with the requirements of paragraph 1910.66(i)(1), in the safe use of the system.

Personal Protection 1910.94(9)(i)

(9) Personal protection. **(i)** All employees working in and around open-surface tank operations must be instructed as to the hazards of their respective jobs, and in the personal protection and first aid procedures applicable to these hazards.

1910.94(9)(vi)

(vi) Respirators shall be used in accordance with ss 1910.134, and persons who may require them shall be trained in their use.

Inspection, Maintenance, and Installation 1910.94(11)(v)

(v) If, in emergencies, such as rescue work, it is necessary to enter a tank which may contain a hazardous atmosphere, suitable respirators, such as self-contained breathing apparatus; hose mask with blower, if there is a possibility of oxygen deficiency; or a gas mask, selected and operated in accordance with paragraph **(b)(9)(vi)** of this section, shall be used. If a contaminant in the tank can cause dermatitis, or be absorbed through the skin, the employee entering the tank shall also wear protective clothing. At least one trained standby employee, with suitable respirator, shall be present in the nearest uncontaminated area. The standby employee must be able to communicate with the employee in the tank and be able to haul him out of the tank with a lifeline if necessary.

Hearing Protection 1910.95(i)(4)

(4) The employer shall provide training in the use and care of all hearing protectors provided to employees.

(1) The employer shall institute a training program for all employees who are exposed to noise at or above an 8–hour time-weighted average of 85 decibels, and shall ensure employee participation in such program.

(2) The training program shall be repeated annually for each employee included in the hearing conservation program. Information provided in the training program shall be updated to be consistent with changes in protective equipment and work processes.

(3) The employer shall ensure that each employee is informed of the following:

(i) The effects of noise on hearing;

(ii) The purpose of hearing protectors, the advantages, disadvantages, and attenuation of various types, and instructions on selection, fitting, use, and care; and

(iii) The purpose of audiometric testing, and an explanation of the test procedures.

Ionizing Radiation 1910.96(i)

(i) All individuals working in or frequenting any portion of a radiation area shall be informed of the occurrence of radioactive materials or radiation in such portions of the radiation area; shall be instructed in the safety problems associated with exposure to such materials or radiation and in precautions or devices to minimize exposure; shall be instructed in the applicable provisions of this section for the protection of employees from exposure to radiation or radioactive materials; and shall be advised of reports of radiation exposure which employees must request pursuant to the regulations in this section.

Testing 1910.96(3)(viii)

(viii) Before placing the system into operation, all employees normally working in the area shall be made acquainted with the signal by actual demonstration at their work locations.

Flammable and Combustible Liquids 1910.106(b)(5)(vi)(v) (2) and (3)

(2) That detailed printed instructions of what to do in flood emergencies are properly posted.

(3) That station operators and other employees depended upon to carry out such instructions are thoroughly informed as to the location and operation of such valves and other equipment necessary to effect these requirements.

Explosives and Blasting Agents 1910.109(d)(3)(i)

(i) Vehicles transporting explosives shall only be driven by and be in the charge of a driver who is familiar with the traffic regulations, state laws, and the provisions of this section.

1910.109(d)(3)(iii)

(iii) Every motor vehicle transporting any quantity of Class A or Class B explosives shall, at all times, be attended by a driver or other attendant of the motor carrier. This attendant shall have been made aware of the class of the explosive material in the vehicle and of its inherent dangers, and shall have been instructed in the measures and procedures to be followed in order to protect the public from those dangers. He shall have been made familiar with the vehicle he is assigned and shall be trained, supplied with the necessary means, and authorized to move the vehicle when required.

1910.109(g)(3)(iii)(a)

(iii)(a) The operator shall be trained in the safe operation of the vehicle together with its mixing, conveying, and related equipment. The employer shall assure that the operator is familiar with the commodities being delivered and the general procedure for handling emergency situations.

1910.109(g)(6)(ii)

(ii) Vehicles transporting blasting agents shall only be driven by and be in charge of a driver in possession of a valid motor vehicle operator's license. Such a person shall also be familiar with the State's vehicle and traffic laws.

Bulk Delivery and Mixing Vehicles
1910.109(3)(d)(iii)

(iii) The operator shall be trained in the safe operation of the vehicle together with its mixing, conveying, and related equipment. He shall be familiar with the commodities being delivered and the general procedure for handling emergency situations.

Storage and Handling of Liquefied Petroleum Gasses
1910.110(b)(16) 1910.110(d)(12)(i)

(16) Instructions. Personnel performing installation, removal, operation, and maintenance work shall be properly trained in such function.

(i) When standard watch service is provided, it shall be extended to the LP-Gas installation and personnel properly trained.

1910.111(b)(13)(ii)

(ii) The employer shall insure that unloading operations are performed by reliable persons properly instructed and given the authority to monitor careful compliance with all applicable procedures.

Process Safety Management of Highly Hazardous Chemicals
1910.119(g)(1)(i) and (ii)

(g) Training. **(1)** Initial training. **(i)** Each employee presently involved in operating a process, and each employee before being involved in operating a newly assigned process, shall be trained in an overview of the process and in the operating procedures as specified in paragraph **(f)** of this section. The training shall include emphasis on the specific safety and health hazards, emergency operations including shutdown, and safe work practices applicable to the employee's job tasks.

(ii) In lieu of initial training for those employees already involved in operating a process on May 26, 1992, an employer may certify in writing that the employee has the required knowledge, skills, and abilities to safely carry out the duties and responsibilities as specified in the operating procedures.

1910.119(g)(2)

(2) Refresher training. Refresher training shall be provided at least every three

years, and more often if necessary, to each employee involved in operating a process to assure that the employee understands and adheres to the current operating procedures of the process. The employer, in consultation with the employees involved in operating the process, shall determine the appropriate frequency of refresher training.

1910.119(g)(3)

(3) Training documentation. The employer shall ascertain that each employee involved in operating a process has received and understood the training required by this paragraph. The employer shall prepare a record which contains the identity of the employee, the date of training, and the means used to verify that the employee understood the training.

Contract Employer Responsibilities
1910.119(h)(3)(i), (ii), (iii), and (iv)

(3) Contract employer responsibilities. **(i)** The contract employer shall assure that each contract employee is trained in the work practices necessary to safely perform his/her job.

(ii) The contract employer shall assure that each contract employee is instructed in the known potential fire, explosion, or toxic release hazards related to his/her job and the process, and the applicable provisions of the emergency action plan.

(iii) The contract employer shall document that each contract employee has received and understood the training required by this paragraph. The contract employer shall prepare a record which contains the identity of the contract employee, the date of training, and the means used to verify that the employee understood the training.

(iv) The contract employer shall assure that each contract employee follows the safety rules of the facility including the safe work practices required by paragraph (f)(4) this section.

Mechanical Integrity 1910.119(j)(3)

(3) Training for process maintenance activities. The employer shall train each employee involved in maintaining the on-going integrity of process equipment in an overview of that process and its hazards and in the procedures applicable to the employee's job tasks to assure that the employee can perform the job tasks in a safe manner.

Hazardous Waste Operations and Emergency Response
1910.120(e)

(e) Training-**(1)** General. All employees working on site (such as but not limited to equipment operators, general laborers and others) exposed to hazardous substances, health hazards, or safety hazards and their supervisors and management responsible for the site shall receive training meeting the requirements of this paragraph before they are permitted to engage in hazardous waste operations that could expose them to hazardous substances, safety or health hazards, and they shall receive review training as specified in this paragraph.

(ii) Employees shall not be permitted to participate in or supervise field activities until they have been trained to a level required by their job function and responsibility.

(2) Elements to be covered. The training shall thoroughly cover the following:

(i) Names of personnel and alternates responsible for the site safety and health;

(ii) Safety, health and other hazards present on the site;

(iii) Use of personal protective equipment;

(iv) Work practices by which the employee can minimize risks from hazards;

(v) Safe use of engineering controls and equipment on the site;

(vi) Medical surveillance requirements, including recognition of symptoms and signs which might indicate overexposure to hazards; and

(vii) The contents of paragraphs **(G)** through **(J)** of the site safety and health plan set forth in paragraph **(b)(4)(ii)** of this section.

(3) Initial training. **(i)** General site workers (such as equipment operators, general laborers and supervisory personnel) engaged in hazardous substance removal or other activities which expose or potentially expose workers to hazardous substances and health hazards shall receive a minimum of 40 hours of instruction off the site, and a minimum of three days actual field experience under the direct supervision of a trained, experienced supervisor.

(ii) Workers on site only occasionally for a specific limited task (such as, but not limited to, ground water monitoring, land surveying, or geo-physical surveying) and who are unlikely to be exposed over permissible exposure limits and published exposure limits shall receive a minimum of 24 hours of instruction off the site, and the minimum of one day actual field experience under the direct supervision of a trained, experienced supervisor.

(iii) Workers regularly on site who work in areas which have been monitored and fully characterized indicating that exposures are under permissible exposure limits and published exposure limits where respirators are not necessary, and the characterization indicates that there are no health hazards or the possibility of an emergency developing, shall receive a minimum of 24 hours of instruction off the site and the minimum of one day actual field experience under the direct supervision of a trained, experience supervisor.

(iv) Workers with 24 hours of training who are covered by paragraphs **(e)(3)(ii)** and **(e)(3)(iii)** of this section, and who become general site workers or who are required to wear respirators, shall have the additional 16 hours and two days of training necessary to total the training specified in paragraph **(e)(3)(i).**

(4) Management and supervisor training. On-site management and supervisors directly responsible for, or who supervise employees engaged in, hazardous waste operations shall receive 40 hours initial training, and three days of supervised field experience (the training may be reduced to 24 hours and one day if the only area of their responsibility is employees covered by paragraphs **(e)(3)(iii)** and at least eight additional hours of specialized training at the time of job assignment on such topics as, but not limited to, the employer's safety and health program and the associated employee training program, personal protective equipment program, spill containment program, and the health hazard monitoring procedure and techniques.

(5) Qualifications for trainers. Trainers shall be qualified to instruct employees about the subject matter that is being presented in training. Such trainers shall have satisfactorily completed a training program for teaching the subjects they are expected to teach, or they shall have the academic credentials and instructional experience necessary for teaching the subjects. Instructors shall demonstrate competent instructional skills and knowledge of the applicable subject matter.

(6) Training certification. Employees and supervisors that have received and successfully completed the training and field experience specified in paragraphs **(e)(1)** through **(e)(4)** of this section shall be certified by their instructor or the head instructor and trained supervisor as having successfully completed the necessary training. A written certificate shall be given to each person so certified. Any person who has not been so certified or who does not meet the requirements of paragraph **(e)(9)** of this section shall be prohibited from engaging in hazardous waste operations.

(7) Emergency response. Employees who are engaged in responding to hazardous emergency situations at hazardous waste clean-up sites that may expose them to hazardous substances shall be trained in how to respond to such expected emergencies.

(8) Refresher training. Employees specified in paragraph **(e)(1)** of this section, and managers and supervisors specified in paragraph **(e)(4)** of this section, shall receive eight hours of refresher training annually on the items specified in paragraph **(e)(2)** and/or **(e)(4)** of this section, any critique of incidents that have occurred in the past year that can serve as training examples of related work, and other relevant topics.

(9) Equivalent training. Employers who can show by documentation or certification that an employee's work experience and/or training has resulted in training equivalent to that training required in paragraphs **(e)(1)** through **(e)(4)** of this section shall not be required to provide the initial training requirements of those paragraphs to such employees. However, certified employees or employees with equivalent training new to a site shall receive appropriate, site specific training before site entry and have appropriate supervised field experience at the new site. Equivalent training includes any academic training or the training that existing employees might have already received from actual hazardous waste site work experience.

However, certified employees or employees with equivalent training new to a site shall receive appropriate site specific training before site entry and have appropriate supervised field experience at the new site. Equivalent training includes any academic training or the training that existing employees might have already received from actual hazardous waste site work experience.

New Technology Programs 1910.120(o)(i)

(i) The employer shall develop and implement procedures for the introduction of effective new technologies and equipment developed for the improved protection of employees working with hazardous waste clean-up operations, and the same shall be implemented as part of the site safety and health program to assure that employee protection is being maintained.

(7) Training program- **(i)** New employees. The employer shall develop and implement a training program, which is part of the employer's safety and health program, for employees exposed to health hazards or hazardous substances at TSD operations to enable the employees to perform their assigned duties and functions in a safe and healthful manner so as not endanger themselves or other employees. The initial training shall be for 24 hours and refresher training shall be for eight hours annually. Employees who have received the initial training required by this paragraph shall be give a written certificate attesting that they have successfully completed the necessary training.

(ii) Current employees. Employers who can show by an employee's previous work experience and/or training that the employee has had training equivalent to the initial training required by this paragraph, shall be considered as meeting the initial training requirements of this paragraph as to that employee. Equivalent training includes the training that existing employees might have already received from actual site work experience. Current employees shall receive eight hours of refresher training annually.

(iii) Trainers. Trainers who teach initial training shall have satisfactorily completed a training course for teaching the subjects they are expected to teach or they shall have the academic credentials and instruction experience necessary to demonstrate a good command of the subject matter of the courses and competent instructional skills.

1910.120(p)(8)(iii)

(iii) Training. **(A)** Training for emergency response employees shall be completed before they are called upon to perform in real emergencies. Such training shall include the elements of the emergency response plan, standard operating procedures the employer has established for the job, the personal protective equipment to be worn and procedures for handling emergency incidents.

1910.120(q)(4)

(4) Skilled support personnel. Personnel, not necessarily an employer's own employees, who are skilled in the operation of certain equipment, such as mechanized earth moving or digging equipment or crane and hoisting equipment, and who are needed temporarily to perform immediate emergency support work that cannot reasonably be performed in a timely fashion by an employer's own employees, and who will be or may be exposed to the hazards at an emergency response scene, are not required to meet the training required in this paragraph for the employer's regular

employees. However, these personnel shall be given an initial briefing at the site prior to their participation in any emergency response. The initial briefing shall include instruction in the wearing of appropriate personal protective equipment, what chemical hazards are involved, and what duties are to be performed. All other appropriate safety and health precautions provided to the employer's own employees shall be used to assure the safety and health of these personnel.

1910.120(q)(5)

(5) Specialist employees. Employees who in the course of their regular job duties, work with and are trained in the hazards of specific hazardous substances, and who will be called upon to provide technical advice or assistance at a hazardous substance release incident to the individual in charge, shall receive training or demonstrate competency in the area of their specialization annually.

1910.120(q)(6)(i)(ii)(iii)(iv)(v)

(6) Training. Training shall be based on the duties and function to be performed by each responder of an emergency response organization. The skill and knowledge levels required for all new responders, those hired after the effective date of this standard, shall be conveyed to them through training before they are permitted to take part in actual emergency operations on an incident. Employees who participate, or are expected to participate in emergency response, shall be given training in accordance with the following paragraphs:

(i) First responder awareness level. First responders at the awareness level are individuals who are likely to witness or discover a hazardous substance release and who have been trained to initiate an emergency response sequence by notifying the proper authorities of the release. They would take no further action beyond notifying the authorities of the release. First responders at the awareness level shall have sufficient training or have had sufficient experience to objectively demonstrate competency in the following areas:

(A) An understanding of what hazardous substances are and the risks associated with them in an incident.

(B) An understanding of the potential outcomes associated with an emergency created when hazardous substances are present.

(C) The ability to recognize the presence of hazardous substances in an emergency.

(D) The ability to identify the hazardous substances, if possible.

(E) An understanding of the role of the first responder awareness individual in the employer's emergency response plan including site security and control and the U.S. Department of Transportation's Emergency Response Guidebook.

(F) The ability to realize the need for additional resources, and to make appropriate notifications to the communication center.

(ii) First responder operations level. First responders at the operations level are individuals who respond to releases or potential releases of hazardous substances as part of the initial response to the site for the purpose of protecting nearby persons, property, or the environment from the effects of the release. They are trained to respond in a defensive fashion without actually trying to stop the release. Their function is to contain the release from a safe distance, keep it from spreading, and prevent exposures. First responders at the operational level shall have received at least eight hours of training or have had sufficient experience to objectively demonstrate competency in the following areas in addition to those listed for the awareness level and the employer shall so certify:

(A) Knowledge of the basis hazard and risk assessment techniques.

(B) Know how to select and use proper personal protective equipment provided to the first responder operational level.

(C) An understanding of basic hazardous materials terms.

(D) Know how to perform basic control, containment and/or confinement operations within the capabilities of the resources and personal protective equipment available with their unit.

(E) Know how to implement basic decontamination procedures.

(F) An understanding of the relevant standard operating procedures and termination procedures.

(iii) Hazardous materials technician. Hazardous materials technicians are individuals who respond to releases or potential releases for the purpose of stopping the release. They assume a more aggressive role than a first responder at the operations level in that they will approach the point of release in order to plug, patch or otherwise stop the release of a hazardous substance. Hazardous materials techniques shall have received at least 24 hours of training equal to the first responder operations level and in addition have competency in the following areas and the employer shall so certify:

(A) Know how to implement the employer's emergency response plan.

(B) Know the classification, identification and verification of known and unknown materials by using field survey instruments and equipment.

(C) Be able to function within an assigned role in the Incident Command System.

(D) Know how to select and use proper specialized chemical personal protective equipment provided to the hazardous materials technician.

(E) Understand hazard and risk assessment techniques.

(F) Be able to perform advance control, containment, and/or confinement operations within the capabilities of the resources and personal protective equipment available with the unit.

(G) Understand and implement decontamination procedures.

(H) Understand termination procedures.

(I) Understand basic chemical and toxicological terminology and behavior.

(iv) Hazardous materials specialist. Hazardous materials specialists are individuals who respond with and provide support to hazardous materials technicians. Their duties parallel those of the hazardous materials technician, however, those duties require a more directed or specific knowledge of the various substances they may be called upon to contain. The hazardous materials specialist would also act as the site liaison with federal, state, local and other government authorities in regards to site activities. Hazardous materials specialists shall have received at least 24 hours of training equal to the technical level and in addition have competency in the following areas and the employer shall so certify:

(A) Know how to implement the local emergency response plan.

(B) Understand classification, identification and verification of known and unknown materials by using advanced survey instruments and equipment.

(C) Know of the state emergency response plan.

(D) Be able to select and use proper specialized chemical personal protective equipment provided to the hazardous materials specialist.

(E) Understand in-depth hazard and risk techniques.

(F) Be able to perform specialized control, containment, and/or confinement operations within the capabilities of the resources and personal protective equipment available.

(G) Be able to determine and implement decontamination procedures.

(H) Have the ability to develop a site safety and control plan.

(I) Understand chemical, radiological and toxicological terminology and behavior.

(v) On scene incident commander. Incident commanders, who will assume control of the incident scene beyond the first responder awareness level shall receive at least 24 hours of training equal to the first responder operations level and in addition have competency in the following areas and the employer shall so certify:

(A) Know and be able to implement the employer's incident command system.

(B) Know how to implement the employer's emergency response plan.

(C) Know and understand the hazards and risks associated with employees working in chemical protective clothing.

(D) Know how to implement the local emergency response plan.

(E) Know of the state emergency response plan and of the Federal Regional Response Team.

(F) Know and understand the importance of decontamination procedures.

1910.120(q)(7)

(7) Trainers. Trainers who teach any of the above training subjects shall have satisfactorily completed a training course for teaching the subjects they are expected to teach, such as the courses offered by the U.S. National Fire Academy, or they shall have the training and/or academic credentials and instructional experience necessary to demonstrate competent instructional skills and a good command of the subject matter of the courses they are to teach.

1910.120(q)(8)

(8) Refresher training. **(i)** Those employees who are trained in accordance with paragraph **(q)(6)** of this section shall receive annual refresher training of sufficient content and duration to maintain their competencies or shall demonstrate competency in those areas at least yearly.

(ii) A statement shall be made of the training or competency and if a statement of competency is made, the employer shall keep a record of the methodology used to demonstrate competency.

Respiratory Protection 1910.134(a)(3) 1910.134(b)(3)

(3) The employee shall use the provided respiratory protection in accordance with instructions and training received.

(3) The user shall be instructed and trained in the proper use of respirators and their limitations.

1910.134(e)(2), (3), (4) and (5)(i)

(2) The correct respirator shall be specified for each job. The respirator type is usually specified in the work procedures by a qualified individual supervising the respiratory protective program. The individual issuing them shall be adequately instructed to insure that the correct respirator is issued. Each respirator permanently assigned to an individual should be durably marked to indicate to whom it was assigned. This mark shall not affect the respirator performance in any way. The date of issuance should be recorded.

(3) Written procedures shall be prepared covering safe use of respirators in dangerous atmospheres that might be encountered in normal operations or in emergencies. Personnel shall be familiar with these procedures and the available respirators.

(4) Respiratory protection is no better than the respirator in use, even though it is worn conscientiously. Frequent random inspections shall be conducted by a qualified individual to assure that respirators are properly selected, used, cleaned and maintained.

(5) For safe use of any respirator, it is essential that the user be properly instructed in its selection, use and maintenance. Both supervisors and workers shall be so instructed by competent persons. Training shall provide the men an opportunity to handle the respirator, have it fitted properly, test its facepiece-to-face seal, wear it in normal air for a long familiarity period and, finally, to wear it in a test atmosphere.

(i) Every respirator wearer shall receive fitting instructions including demonstrations and practice in how the respirator should be worn, how to adjust it, and how to determine if it fits properly. Respirators shall not be worn when conditions prevent a good face seal. Such conditions may be a growth of beard, sideburns, a skull cap that projects under the facepiece or temple pieces on glasses. Also, the absence of one or both dentures can seriously affect the fit of a facepiece. The worker's diligence in observing these factors shall be evaluated by periodic check. To assure proper protection, the facepiece fit shall be checked by the wearer each time he puts on the respirator. This may be done by following the manufacturer's facepiece fitting instructions.

Temporary Labor Camps 1910.142(k)(1) and (2)

(1) Adequate first aid facilities approved by a health authority shall be maintained and made available in every labor camp for the emergency treatment of injured persons.

(2) Such facilities shall be in charge of a person trained to administer first aid and shall be readily accessible for use at all times.

Specifications for Accident Prevention Signs and Tags 1910.145(c)(1)(ii), (2)(ii) and (3)

(1)(ii) All employees shall be instructed that danger signs indicate immediate danger and that special precautions are necessary.

(2)(ii) All employees shall be instructed that caution signs indicate a possible hazard against which proper precautions should be taken.

(3) Safety instruction signs. Safety instruction signs shall be used where there is a need for general instructions and suggestions relative to safety measures.

Medical Services and First Aid 1910.151(a) and (b)

(a) The employer shall ensure the ready availability of medical personnel for advice and consultation on matters of plant health.

(b) In the absence of an infirmary, clinic or hospital in near proximity to the workplace which is used for the treatment of all injured employees, a person or persons shall be adequately trained to render first aid. First aid supplies approved by the consulting physician shall be readily available.

The Control of Hazardous Energy (lockout/tagout) 1910.147(a)(3)(ii)

(ii) When other standards in this part require the use of lockout or tagout, they shall be used and supplemented by the procedural and training requirements of this section.

(4) Energy control procedure. **(i)** Procedures shall be developed, documented and utilized for the control of potentially hazardous energy when employees are engaged in the activities covered by this section.

(d) Where tagout is used for energy control, the periodic inspection shall include a review between the inspector and each authorized and affected employee of that employee's responsibilities under the energy control procedure being inspected and the elements set forth in paragraph **(c)(7)(ii)** of this section.

(7) Training and communication. **(i)** The employer shall provide training to ensure that the purpose and function of the energy control program are understood by employees and that the knowledge and skills required for the safe application, usage and removal of energy controls are required by employees. The training shall include the following:

(A) Each authorized employee shall receive training in the recognition of applicable hazardous energy sources, the type and magnitude of the energy available in the workplace and the methods and means necessary for energy isolation and control.

(B) Each affected employee shall be instructed in the purpose and use of the energy control procedure.

(C) All other employees whose work operations are or may be in an area where energy control procedures may be utilized shall be instructed about the procedure and about the prohibition relating to attempts to restart or reenergize machines or equipment which are locked out or tagged out.

(ii) When tagout systems are used employees shall also be trained in the following limitations of tags:

(A) Tags are essentially warning devices affixed to energy isolating devices and do not provide the physical restraint on those devices and do not provide the physical restraint on those devices that is provided by a lock.

(B) When a tag is attached to an energy isolating means, it is not to be removed without authorization of the authorized person for it and it is never to be bypassed, ignored or otherwise defeated.

(C) Tags must be legible and understandable by all authorized employees, affected employees and all other employees whose work operations are or may be in the area in order to be effective.

(D) Tags and their means of attachment must be made of materials which will withstand the environmental conditions encountered in the workplace.

(E) Tags may evoke a false sense of security and their meaning needs to be understood as part of the overall energy control program.

(F) Tags must be securely attached to energy isolating devices so that they cannot be inadvertently or accidentally detached during use.

(iii) Employee retraining.

(A) Retraining shall be provided for all authorized and affected employees whenever there is a change in their job assignments, a change in machines, equipment or processes that present a new hazard or when there is a change in the energy control procedures.

(B) Additional retraining shall also be conducted whenever a periodic inspection under paragraph (c)(6) of this section reveals or whenever the employer has reason to believe that there are deviations from or inadequacies in the employee's knowledge or use of the energy control procedures.

(C) The retraining shall reestablish employee proficiency and introduce new or revised control methods and procedures, as necessary.

(iv) The employer shall certify that employee training has been accomplished and is being kept up to date. The certification shall contain each employee's name and dates of training.

(8) Energy isolation. Implementation of lockout or the tagout system shall be performed only by authorized employees.

Lockout or Tagout Devices Removed 1910.147(e)(3)

(3) Lockout or tagout devices removal. Each lockout or tagout device shall be removed from each energy isolating device by the employee who applied the device.

Exception to paragraph (e)(3): When the authorized employee who applied the lockout or tagout device is not available to remove it, that device may be removed under the direction of the employer provided that specific procedures and training for such removal have been developed, documented and incorporated into the employer's energy control program. The employer shall demonstrate that the specific procedure provides equivalent safety to the removal of the device by the authorized employee who applied it.

Outside Personnel 1910.147(f)(2)(i)

(i) The on-site employer shall ensure that his/her personnel understand and comply with restrictions and prohibitions of the outside employer's energy control procedures.

Medical Services and First Aid 1910.151(b)

(b) In the absence of an infirmary, clinic or hospital in near proximity to the workplace which is used for the treatment of all injured employees, a person or persons shall be adequately trained to render first aid. First aid supplies approved by the consulting physician shall be readily available.

Fire Protection 1910.155(iv)(41)

(41) Training means the process of making proficient through instruction and hands on practice in the operation of equipment including respiratory protection equipment that is expected to be used and in the performance of assigned duties.

Fire Brigades 1910.156(b)(1)

(1) Organization—**(1)** Organizational statement. The employer shall prepare and maintain a statement or written policy which establishes the existence of a fire brigade; the basic organizational structure; the type, amount and frequency of training to be provided to fire brigade members; the expected number of members in the fire brigade; and the functions that the fire brigade is to perform at the workplace. The organizational statement shall be available for inspection by the Assistant Secretary and by employees or their designated representatives.

Training and Education 1910.156(c)(1)(2)(3)(4)

(c) Training and education. **(1)** Organizational statement. The employer shall provide training and education for all fire brigade members commensurate with those duties and functions that fire brigade members are expected to perform. Such training and education shall be provided to fire brigade members before they perform fire brigade emergency activities. Fire brigade leaders and training instructors shall be provided with training and education which is more comprehensive than that provided to the general membership of the fire brigade.

(2) The employer shall assure that training and education is conducted frequently enough to assure that each member of the fire brigade is able to perform the member's assigned duties and functions satisfactorily and in a safe manner so as not to endanger fire brigade members or other employees. All fire brigade members shall be provided with training at least annually. In addition, fire brigade members who are expected to perform interior structural fire fighting shall be provided with an education session or training at least quarterly.

(3) The quality of the training and education program for fire brigade members shall be similar to those conducted by such fire training schools as the Maryland Fire and Rescue Institute; Iowa Fire Service Extension; West Virginia Fire Service Extension; Georgia Fire Academy; New York State Department Fire Prevention and Control; Louisiana State University Firemen Training Program; or Washington State's Fire Service Training Commission for Vocational Education. (For example, for the oil refinery industry with its unique hazards, the training and education program for those fire brigade members shall be similar to those conducted by Texas A & M University, Lamar University, Reno Fire School or the Delaware State Fire School.)

(4) The employer shall inform fire brigade members about special hazards such as storage and use of flammable liquids and gases, toxic chemicals, radioactive sources and water reactive substances, to which they may be exposed during fire and other emergencies. The fire brigade members shall also be advised of any changes that occur in relation to the special hazards. The employer shall develop and make available for inspection by fire brigade members, written procedures that describe the actions to be taken in situations involving the special hazards and shall include these in the training and education program.

Portable Fire Extinguishers 1910.157(g)

(g) Training and education.

(1) Where the employer has provided portable fire extinguishers for employee use in the workplace, the employer shall also provide an education program to familiarize employees with the general principles of fire extinguisher use and the hazards involved with incipient stage fire fighting.

(2) The employer shall provide the education required in paragraph (g)(1) of this section upon initial employment and at least annually thereafter.

(3) The employer shall provide employees who have been designated to use fire fighting equipment as part of an emergency action plan with training in the use of the appropriate equipment.

(4) The employer shall provide the training required in paragraph (g)(3) of this section upon initial assignment to the designated group of employees and at least annually thereafter.

Fixed Fire Extinguishers 1910.159(e)(2)(vi)

(vi) The employer shall designate trained persons to conduct all inspections required under this section.

Fixed Extinguishing Systems 1910.160(b)(10)

(10) The employer shall train employees designated to inspect, maintain, operate or repair fixed extinguishing systems and annually review their training to keep them up to date in the functions they are to perform.

Fire Detection Systems 1910.164(c)(4)

(4) The employer shall assure that the servicing, maintenance and testing of fire detection systems, including cleaning and necessary sensitivity adjustments are performed by a trained person knowledgeable in the operations and functions of the system.

Servicing of Multi-Piece and Single-Piece Rim Wheels 1910.177(c)(1)(2) and (3).

(Includes single piece wheels per Federal Register of February 3, 1984 **(pp.4338–4352) but not automobile or truck tires marked LT.)**

(c) Employee training.

(1) The employer shall provide a training program to train all employees who service rim wheels in the hazards involved in servicing those rim wheels and the safety procedures to be followed.

(i) The employer shall assure that no employee services any rim wheel unless the employee has been trained and instructed in correct procedures of servicing the rim type being serviced and in the safe operating procedures described in paragraphs (f) and (g) of this section.

(ii) Information to be used in the training program shall include at a minimum, the applicable data contained in the charts, rim manuals and the contents of this standard.

(iii) Where an employer knows or has reason to believe that any of his employees is unable to read and understand the charts or rim manual, the employer shall assure that the employee is instructed concerning the contents of the charts and rim manual in a manner which the employee is able to understand.

(2) The employer shall assure that each employee demonstrates and maintains the ability to service multi-piece rim wheels safely, including performance of the following tasks:

(i) Demounting of tires (including deflation);

(ii) Inspection and identification of rim wheel components;

(iii) Mounting of tires (including inflation within a restraining device or other safeguard required by this section);

(iv) Use of the restraining device or barrier and other equipment required by this section;

(v) Handling of rim wheels;

(vi) Inflation of tire when a single piece rim wheel is mounted on a vehicle; and

(vii) An understanding of the necessity of standing outside the trajectory both during the inflation of the tire and during inspection of the rim wheel following inflation; and

(viii) Installation and removal of rim wheels.

(3) The employer shall evaluate each employee's ability to perform these tasks and to service rim wheels safely and shall provide additional training as necessary to assure that each employee maintains his or her proficiency.

1910.177(f)

(f) Safe operating procedure—multi-piece rim wheels. The employer shall establish a safe operating procedure for servicing multi-piece rim wheels and shall assure that employees are instructed in and follow that procedure. The procedure shall include at least the following elements:

(1) Tires shall be completely deflated before demounting by removal of the valve core.

(2) Tires shall be completely deflated by removing the valve core before a rim wheel is removed from the axle in either of the following situations:

(i) When the tire has been driven under inflated at 80% or less of its recommended pressure, or

(ii) When there is obvious or suspected damage to the tire or wheel components.

(3) Rubber lubricant shall be applied to bead and rim mating surfaces during assembly of the wheel and inflation of the tire, unless the tire or wheel manufacturer recommends against it.

(4) If a tire on a vehicle is under inflated but has more than 80% of the recommended pressure, the tire may be inflated while the rim wheel is on the vehicle provided remote control inflation equipment is used and no employees remain in the trajectory during inflation.

(5) Tires shall be inflated outside a restraining device only to a pressure sufficient to force the tire bead onto the rim ledge and create an airtight seal with the tire and bead.

(6) Whenever a rim wheel is in a restraining device the employee shall not rest or lean any part of his body or equipment on or against the restraining device.

(7) After tire inflation, the tire and wheel components shall be inspected while still within the restraining device to make sure that they are properly seated and locked. If further adjustment to the tire or wheel components is necessary, the tire shall be deflated by removal of the valve core before the adjustment is made.

(8) No attempt shall be made to correct the seating of side and lock rings by hammering, striking or forcing the components while the tire is pressurized.

(9) Cracked, broken, bent or otherwise damaged rim components shall not be reworked, welded, brazed or otherwise heated.

(10) Whenever multi-piece rim wheels are being handled, employees shall stay out of the trajectory unless the employer can demonstrate that performance of the servicing makes the employee's presence in the trajectory necessary.

(11) No heat shall be applied to a multi-piece wheel or wheel component.

1910.177(g)

(g) Safe operating procedure—single piece rim wheels. The employer shall establish a safe operating procedure for servicing single piece rim wheels and shall assure that employees are instructed in and follow that procedure. The procedure shall include at least the following elements:

(1) Tires shall be completely deflated by removal of the valve core before demounting.

(2) Mounting and demounting of the tire shall be done only from the narrow ledge side of the wheel. Care shall be taken to avoid damaging the tire beads while mounting tires on wheels. Tires shall be mounted only on compatible wheels of matching bead diameter and width.

(3) Nonflammable rubber lubricant shall be applied to bead and wheel mating surfaces before assembly of the rim wheel, unless the tire or wheel manufacturer recommends against the use of any rubber lubricant.

(4) If a tire changing machine is used, the tire shall be inflated only to the minimum pressure necessary to force the tire bead onto the rim ledge while on the tire changing machine.

(5) If a bead expander is used, it shall be removed before the valve core is installed and as soon as the rim wheel becomes airtight (the tire bead slips onto the bead seat).

(6) Tires may be inflated only when contained within a restraining device, positioned behind a barrier or bolted on the vehicle with the lug nuts fully tightened.

(7) Tires shall not be inflated when any flat, solid surface is in the trajectory and within one foot of the sidewall.

(8) Employees shall stay out of the trajectory when inflating a tire.

(9) Tires shall not be inflated to more than the inflation pressure stamped in the sidewall unless higher pressure is recommended by the manufacturer.

(10) Tires shall not be inflated above the maximum pressure recommended by the manufacturer to seat the tire bead firmly against the rim flange.

(11) No heat shall be applied to a single piece wheel.

(12) Cracked, broken, bent or otherwise damaged wheels shall not be reworked, welded, brazed or otherwise heated.

Powered Industrial Trucks 1910.178

Operator training. Only trained and authorized operators shall be permitted to operate a powered industrial truck. Methods shall be devised to train operators in the safe operation of powered industrial trucks.

Moving the Load 1910.179(n)(3)(ix)

(ix) When two or more cranes are used to lift a load one qualified responsible person shall be in charge of the operation. He shall analyze the operation and instruct all personnel involved in the proper positioning, rigging of the load and the movements to be made.

1910.179(o)(3)

(3) Fire extinguishers. The employer shall insure that operators are familiar with the operation and care of fire extinguishers provided.

Crawler Locomotive and Truck Cranes 1910.180(i)(5)(ii)

(ii) Operating and maintenance personnel shall be made familiar with the use and care of the fire extinguishers provided.

Mechanical Power Presses 1910.217(e)(3)

(3) Training of maintenance personnel. It shall be the responsibility of the employer to insure the original and continuing competence of personnel caring for, inspecting and maintaining power presses.

1910.217(f)(2)

(2) Instruction to operators. The employer shall train and instruct the operator in the safe method of work before starting work on any operation covered by this section. The employer shall insure by adequate supervision that correct operating procedures are being followed.

Mechanical Power Presses Instruction to Operators 1910.217(9)(e)(2)

(2) Instruction to operators. The employer shall train and instruct the operator in the safe method of work before starting work on any operation covered by this section. The employer shall insure by adequate supervision that correct operating procedures are being followed.

Training of Maintenance Personnel 1910.217(9)(e)(3)

(3) Training of maintenance personnel. It shall be the responsibility of the employer to insure the original and continuing competence of personnel caring for, inspecting and maintaining power presses.

Operator Training 1910.217(h)(13)(i)

(13) Operator training. **(i)** The operator training required by paragraph **(f)(2)** of this section shall be provided to the employee before the employee initially operates the press and as needed to maintain competence, but not less than annually thereafter. It shall include instruction relative to the following items for presses used in the PSI mode.

(A) The manufacturer's recommended test procedures for checking operation of the presence sensing device. This shall include the use of the test rod required by paragraph **(h)(10)(i)** of this section.

(B) The safety distance required.

(C) The operation, function and performance of the PSDI mode.

(D) The requirements for hand tools that may be used in the PSDI mode.

(E) The severe consequences that can result if he or she attempts to circumvent or by-pass any of the safeguard or operating functions of the PSDI system.

(ii) The employer shall certify that employees have been trained by preparing a certification record which includes the identity of the person trained, the signature of the employer or the person who conducted the training, and the date the training was completed. The certification record shall be prepared at the completion of training and shall be maintained on file for the duration of the employee's employment. The certification record shall be made available upon request to the Assistant Secretary for Occupational Safety and Health.

Forging Machine 1910.218(a)(2) (iii)

(2) Inspection and maintenance. It shall be the responsibility of the employer to maintain all forge shop equipment in a condition which will ensure continued safe operation. This responsibility includes:

(iii) Training personnel for the proper inspection and maintenance of forging machinery and equipment.

Oxygen-fuel Gas Welding and Cutting 1910.253(a)(1)(4)

(4) Personnel. Workmen in charge of the oxygen or fuel-gas supply equipment, including generators and oxygen or fuel-gas distribution piping systems shall be instructed by their employers for this important work before being left in charge. Rules and instructions covering the operation and maintenance of oxygen or fuel-gas supply equipment including generators and oxygen or fuel-gas distribution piping systems shall be readily available.

Arc Welding and Cutting 1910.154(a)(3)

(3) Instruction. Workmen designated to operate arc welding equipment shall have been properly instructed and qualified to operate such equipment as specified in paragraph **(d)** of this section.

Resistance Welding 1910.255(a)(3)

(3) Personnel. Workmen designated to operate resistance welding equipment shall have been properly instructed and judged competent to operate such equipment.

Pulp, Paper and Paperboard Mills 1910.261(h)(3)(ii)

(ii) Gas masks capable of absorbing chlorine shall be supplied, conveniently placed and regularly inspected and workers who may be exposed to chlorine has shall be instructed in their use.

Laundry Machinery and Operating Rules 1910.264(d)(1)(v)

(v) Instruction of employees. Employees shall be properly instructed as to the hazards of their work and be instructed in safe practices, by bulletins, printed rules and verbal instructions.

Sawmills 1910.265(c)(30)(x)

(x) Lift trucks. Lift trucks shall be designed, constructed, maintained and operated in accordance with the requirements of 1910.178.

Pulpwood Logging 1910.266(c)(5) (i) through (xi)

(i) Chain saw operators shall be instructed to inspect the saws daily to assure that all handles and guards are in place and tight, that all controls function properly and that the muffler is operative.

(ii) Chain saw operators shall be instructed to follow manufacturer's instructions as to operation and adjustment.

(iii) Chain saw operators shall be instructed to fuel the saw only in safe areas and not under conditions conducive to fire such as near men smoking, hot engine, etc.

(iv) Chain saw operators shall be instructed to hold the saw with both hands during operation.

(v) Chain saw operators shall be instructed to start the saw at least 10 feet away from fueling area.

(vi) Chain saw operators shall be instructed to start the saw only on the ground or when otherwise firmly supported.

(vii) Chain saw operators shall be instructed to be certain of footing and to clear away brush which might interfere before starting to cut.

(viii) Chain saw operators shall be instructed not to use engine fuel for starting fires or as a cleaning solvent.

(ix) Chain saw operators shall be instructed to shut off the saw when carrying it for a distance greater than from tree to tree or in hazardous conditions such a slippery surfaces or heavy underbrush. The saw shall be at idle speed when carried short distances.

(x) Chain saw operators shall be instructed to carry the saw in a manner to prevent contact with the chain and muffler.

(xi) Chain saw operators shall be instructed not to use the saw to cut directly overhead or at a distance that would require the operator to relinquish a safe grip on the saw.

1910.266(c)(6)(i) through (xxi)

(6) Stationary and mobile equipment operation. **(i)** Equipment operators shall be instructed as to the manufacturers recommendations for equipment operation, maintenance, safe practices and site operating procedures.

(ii) Equipment shall be kept free of flammable material.

(iii) Equipment shall be kept free of any material which might contribute to slipping and falling.

(iv) Engine of equipment shall be shut down during fueling, servicing and repairs except where operation is required for adjustment.

(v) Equipment shall be inspected for evidence of failure or incipient failure.

(vi) The equipment operator shall be instructed to walk completely around machine and assure that no obstacles or personnel are in the area before startup.

(vii) The equipment operator shall be instructed to start and operate equipment only from the operator's station or from safe area recommended by the manufacturer.

(viii) Seat belt shall be provided on mobile equipment.

(ix) The equipment operator shall be instructed to check all controls for proper function and response before starting working cycle.

(x) The equipment operator shall be instructed to ground or secure all movable elements when not in use.

(xi) The equipment operator shall be advised of the load capacity and operating speed of the equipment.

(xii) The equipment operator shall be advised of the stability limitation of the equipment.

(xiii) The equipment operator shall be instructed to maintain adequate distance from other equipment and personnel.

(xiv) The equipment operator shall be instructed not to operate movable elements (boom, grapple, load, etc.) close to or over personnel.

(xv) The equipment operator shall be instructed to signal his intention before operation when personnel are in or near the working area.

(xvi) The equipment operator shall be instructed to dismount and stand clear for all loading and unloading of his mobile vehicle by other mobile equipment. The dismounted operator shall be visible to loader operator.

(xvii) The equipment operator shall be instructed to operate equipment in a manner that will not place undue shock loads on wire rope.

(xviii) The equipment operator shall be instructed not to permit riders or observers on the machine unless approved seating and protection is provided.

(xix) The equipment operator shall be instructed to shut down the engine when the equipment is stopped, apply brake locks and ground moving elements before he dismounts.

(xx) The equipment operator shall be instructed, when any equipment is transported from one job location to another, to transport it on a vehicle of sufficient rated capacity and the equipment shall be properly secured during transit.

1910.266(c)(7)

(7) Explosives. Only trained and experienced personnel shall handle or use explosives. Usage shall comply with the requirements of 1910.109.

1910.266(e)(2)(i) and (ii)

(i) The feller shall be instructed to plan retreat path and clear path as necessary before cut is started.

(ii) The feller shall be instructed to appraise situation for dead limbs, the lean of tree to be cut, wind conditions, location of other trees and other hazards and exercise proper precautions before cut is started.

Telecommunications 1910.268(b)(2)(i)

(i) Employees assigned to work with storage batteries shall be instructed in emergency procedures such as dealing with accidental acid spills.

1910.268(c)

(c) Training. Employers shall provide training in the various precautions and safe practices described in this section and shall insure that employees do not engage in the activities to which this section applies until such employees have received proper training in the various precautions and safe practices required by this section. However, where the employer can demonstrate that an employee is already trained in the precautions and safe practices required by this section prior to his employment, training need not be provided to that employee in accordance with this section. Where training is required, it shall consist of on-the-job training or classroom type training or a combination of both. The employer shall certify that employees have been trained by preparing a certification record which includes the identity of the person trained, the signature of the employer or the person who conducted the training, and the date the training was completed. The certification record shall be prepared at the completion of training and shall be maintained on file for the duration of the employee's employment. The certification record shall be made available upon request to the Assistant Secretary for Occupational Safety and Health. Such training shall, where appropriate, include the following subjects:

(1) Recognition and avoidance of dangers relating to encounters with harmful substances and animal, insect or plant life;

(2) Procedures to be followed in emergency situations; and,

(3) First aid training, including instruction in artificial respiration.

Derrick Trucks 1910.268(j)(4)(iv)(D)

(D) Only persons trained in the operation of the derrick shall be permitted to operate the derrick.

Cable Fault Locating 1910.268(l)(1)

(1) Cable fault locating and testing. **(1)** Employees involved in using high voltages to locate trouble or test cables shall be instructed in the precautions necessary for their own safety and the safety of other employees.

Guarding Manholes 1910.268(o)(1)(ii)

(ii) While work is being performed in the manhole, a person with basic first aid training shall be immediately available to render assistance if there is cause for believing that a safety hazard exists and if the requirements contained in paragraphs **(d)(1)** and **(o)(1)(i)** of this section do not adequately protect the employee(s).

Joint Power and Telecommunication Manholes 1910.268(o)(3)

(3) Joint power and telecommunication manholes. While work is being performed in a manhole occupied jointly by an electric utility and a telecommunication utility, and employee with basic first aid training shall be available in the immediate vicinity to render emergency assistance as may be required. The employee whose presence is required in the immediate vicinity for the purposes of rendering emergency assistance is not to be precluded from occasionally entering a manhole to provide assistance other than in an emergency. The requirement of this paragraph **(o)(3)** does not preclude a qualified employee, working alone, from entering for brief periods of time, a manhole where energized cables or equipment are in service, for the purpose of inspection, housekeeping, taking readings, or similar work if such work can be performed safely.

Tree Trimming—Electrical Hazards 1910.268(q)(1)(ii)

(ii) Employees engaged in line-clearing operations shall be instructed that:

(A) A direct contact is made when any part of the body touches or contacts an energized conductor or other energized electrical fixture or apparatus.

(B) An indirect contact is made when any part of the body touches any object in contact with an energized electrical conductor or other energized fixture or apparatus.

(C) An indirect contact can be made through conductive tools, tree branches, trucks, equipment, or other objects, as a result of communications wires, cables, fences, or guy wires being accidentally energized.

(D) Electric shock will occur when an employee, by either direct or indirect contact with an energized conductor, energized tree limb, tool equipment, or other object, provides a path for the flow of electricity to a grounded object or to the ground itself. Simultaneous contact with two energized conductors will also cause electric shock which may result in serious or fatal injury.

1910.268(q)(2)(ii)

(ii) Only qualified employees or trainees familiar with the special techniques and hazards involved in line clearance shall be permitted to perform the work if it is found that an electrical hazard exists.

1910.268(q)(2)(iii)

(iii) During all tree working operations aloft where an electrical hazard of more than 750V exists, there shall be a second employee or trainee qualified in line clearance tree trimming within normal voice communication.

Grain Handling Facilities 1910.272(e)

(e) Training. **(1)** The employer shall provide training to employees at least annually and when changes in job assignment will expose them to new hazards. Current employees and new employees, prior to starting work, shall be trained in at least the following:

(i) General safety precautions associated with the facility, including recognition and preventive measures for the hazards related to dust accumulations and common ignition sources such as smoking; and,

(ii) Specific procedures and safety practices applicable to their job tasks including but not limited to, cleaning procedures for grinding equipment, clearing procedures for choked legs, housekeeping procedures, hot work procedures, preventive maintenance procedures and lock-out/tag-out procedures.

(2) Employees assigned special tasks, such as bin entry and handling of flammable or toxic substances, shall be provided training to perform these tasks safely.

Entry into Bins, Silo and Tanks 1910.272(g)(5)

(5) The employee acting as observer shall be trained in rescue procedures, including notification methods for obtaining additional assistance.

Contractors 1910.272(h)(2)

(2) The employer shall explain the applicable provisions of the emergency action plan to contractors.

Electrical Safety-Related Work Practices Content of Training 1910.332(b)(1)

(b) Content of training. **(1)** Practices addressed in this standard. Employees shall be trained in and familiar with the safety related work practices required by ss 1910.331 through 1910.335 that pertain to their respective job assignments.

Qualifications of Dive Team 1910.410(a)(1) through (4)

(1) Each dive team member shall have the experience or training necessary to perform assigned tasks in a safe and healthful manner.

(2) Each dive team member shall have experience or training in the following:

(i) The use of tools, equipment and systems relevant to assigned tasks;

(ii) Techniques of the assigned diving mode; and

(iii) Diving operations and emergency procedures.

(3) All dive team members shall be trained in cardiopulmonary resuscitation and first aid (American Red Cross standard course or equivalent).

(4) Dive team members who are exposed to or control the exposure of others to hyperbaric conditions shall be trained in diving related physics and physiology.

1910.410(b)(1)

(1) Each dive team member shall be assigned tasks in accordance with the employee's experience or training, except that limited additional tasks may be assigned to an employee undergoing training provided that these tasks are performed under the direct supervision of an experienced dive team member.

1910.410(c)(2)

(2) The designated person in charge shall have experience and training in the conduct of the assigned diving operation.

Asbestos 1910.1001(j)(5)

(5) Employee information and training. **(i)** The employer shall institute a training program for all employees who are exposed to airborne concentrations of asbestos, tremolite, anthophylite, actinlite or a combination of these minerals at or above the action level and ensure their participation in the program.

(ii) Training shall be provided prior to or at the time of initial assignment and at least annually thereafter.

(iii) The training program shall be conducted in a manner which the employee is able to understand. The employer shall ensure that each employee is informed of the following:

(A) The health effect associated with asbestos, tremolite, anthophylite or actinolite exposure;

(B) The relationship between smoking and exposure to asbestos, tremolite, anthophylite and actinolite in producing lung cancer;

(C) The quantity, location, manner of use, release and storage of asbestos, tremolite, anthophylite or actinolite and the specific nature of operations which could result in exposure to asbestos, tremolite, anthophylite, or actinolite;

(D) The engineering controls and work practices associated with the employee's job assignment;

(E) The specific procedures implemented to protect employees from exposure to asbestos, tremolite, anthophylite or actinolite, such as appropriate work practices, emergency and clean up procedures and personal protective equipment to be used;

(F) The purpose, proper use and limitations of respirators and protective clothing;

(G) The purpose and a description of the medical surveillance program required by paragraph **(1)** of this section;

4–Nitrobiphenyl 1910.1003(e)(5)(i) and (ii)

(i) Each employee, prior to being authorized to enter a regulated area, shall receive a training and indoctrination program including, but not necessarily limited to:

Alpha-Napthylamine 1910.1004(e)(5)(i) and (ii)

(a) The nature of the carcinogenic hazards of N-Nitrobiphenyl, and others listed below, including local and systemic toxicity;

Methyl Chloromethyl Ether 1910.1006(e)(5)(i) and (ii)

(b) The specific nature of the operation involving 4–Nitrobiphenyl which could result in exposure;

Dichlorobenzidine (and its salts) 1910.1007(e)(5)(i) and (ii)

(c) The purpose for and application of the medical surveillance program, including, as appropriate, methods of self-examination;

Bis-Chloromethyl Ether 1910.1008(e)(5)(i) and (ii)

(d) The purpose for and application of decontamination practices and purposes;

Beta-Naphthylamine 1910.1009(e)(5)(i) and (ii)

(e) The purpose for and significance of emergency practices and procedures;

Benzidine 1910.1010(e)(5)(i) and (ii)

(f) The employee's specific role in emergency procedures;

4–Aminodiphenyl 1910.1011(e)(5)(i) and (ii)

(g) Specific information to aid the employee in recognition and evaluation of conditions and situations which may result in the release of 4–Nitrobiphenyl;

Ethyleneimine 1910.1012(e)(5)(i) and (ii)

(h) The purpose for and application of specific first aid procedures and practices;

Beta-Propiolactone 1910.1013(e)(5)(i) and (ii)

(i) A review of this section at the employee's first training and indoctrination program and annually thereafter.

2–Acetylaminofluorene 1910.1014(e)(5)(i) and (ii)

(ii) Specific emergency procedures shall be prescribed and posted and employees shall be familiarized with their terms and rehearsed in their application.

4–Dimethylaminoazobenzene 1910.1015(e)(5)(i) and (ii)

N-Nitrosodimethylamine 1910.1016(e)(5)(i) and (ii)

Vinyl Chloride 1910.107(j)(1)(i) through (ix)

(j) Training. Each employee engaged in vinyl chloride or polyvinyl chloride operations shall be provided training in a program relating to the hazards of vinyl chloride and precautions for its safe use.

(1) The program shall include:

(i) The nature of the health hazard from chronic exposure to vinyl chloride including specifically the carcinogenic hazard;

(ii) The specific nature of operations which could result in exposure to vinyl chloride in excess of the permissible limit and necessary protective steps;

(iii) The purpose for, proper use and limitations of respiratory protective devices;

(iv) The fire hazard and acute toxicity of vinyl chloride and the necessary protective steps;

(v) The purpose for and a description of the monitoring program;

(vi) The purpose for, and a description of, the medical surveillance program;

(vii) Emergency procedures;

(viii) Specific information to aid the employee in recognition of conditions which may result in the release of vinyl chloride; and

(ix) A review of this standard at the employee's first training and indoctrination program and annually thereafter.

Inorganic Arsenic 1910.1018(o)(1) and (2)

(1) Training program.

(i) The employer shall institute a training program for all employees who are subject to exposure to inorganic arsenic above the action level without regard to respirator use or for whom there is a possibility of skin or eye irritation from inorganic arsenic. The employer shall assure that those employees participate in the training program.

(ii) The training program shall be provided by October 1, 1978 for employees covered by this provision, at the time of initial assignment for those subsequently covered by this provision and shall be repeated at least quarterly for employees who have optional use of respirators and at least annually for other covered employees thereafter and the employer shall assure that each employee is informed of the following:

(A) The information contained in Appendix A;

(B) The quantity, location, manner of use, storage, sources of exposure, and the specific nature of operations which could result in exposure to inorganic arsenic as well as any necessary protective steps;

(C) The purpose, proper use and limitation of respirators;

(D) The purpose and a description of the medical surveillance program as required by paragraph **(n)** of this section;

(E) The engineering controls and work practices associated with the employee's job assignment; and

(F) A review of this standard.

(2) Access to training materials.

(i) The employer shall make readily available to all affected employees a copy of this standard and its appendixes.

(ii) The employer shall provide, upon request, all materials relating to the employee information and training program to the Assistant Secretary and the Director.

Lead 1910.1025(l)(1) and (2)

(1)

(i) Each employer who has a workplace in which there is a potential exposure to airborne lead at any level shall inform employees of the content of Appendixes A and B of this regulation.

(ii) The employer shall institute a training program for and assure the participation of all employees who are subject to exposure to lead at or above the action level or for whom the possibility of skin or eye irritation exists.

(iii) The employer shall provide initial training by 180 days from the effective date. (Editor's Note: OSHA's lead standard became effective February 1, 1979 for those employees covered by paragraph (l)(1)(ii) on the standard's effective date and prior to the time of initial job assignment for those employees subsequently covered by this paragraph.)

(iv) The training program shall be repeated at least annually for each employee.

(v) The employer shall assure that each employee is informed of the following:

(A) The content of this standard and its appendixes;

(B) The specific nature of the operations which could result in exposure to lead above the action level;

(C) The purpose, proper selection, fitting, use and limitations of respirators;

(D) The purpose and a description of the medical surveillance program and the medical removal protection program including information concerning the adverse health effects associated with excessive exposure to lead (with particular attention to the adverse reproductive effects on both males and females);

(E) The engineering controls and work practices associated with the employee's job assignment;

(F) The contents of any compliance plan in effect; and

(G) Instructions to employees that chelating agents should not routinely be used to remove lead from their bodies and should not be used at all except under the direction of a licensed physician;

(2) Access to information and training materials.

(i) The employer shall make readily available to all affected employees a copy of this standard and its appendixes.

(ii) The employer shall provide, upon request, all materials relating to the employee information and training program to the Assistant Secretary and the Director.

(iii) In addition to the information required by paragraph **(l)(1)(v)**, the employer shall include as part of the training program, and shall distribute to employees, any materials pertaining to the Occupational Safety and Health Act, the regulations issued pursuant to that Act, and this lead standard, which are made available to the employer by the Assistant Secretary.

Coke Oven Emissions 1910.1029(k)(1) and (2)

(1) Training program.

(i) The employer shall institute a training program for employees who are employed in the regulated area and shall assure their participation.

(ii) The training program shall be provided as of January 27, 1977 for employees who are employed in the regulated area at that time or at the time of initial assignment to a regulated area.

(iii) The training program shall be provided at least annually for all employees who are employed in the regulated area, except that training regarding the occupational safety and health hazards associated with exposure to coke oven emissions and the purpose, proper use and limitations of respiratory protective devices shall be provided at least quarterly until January 20, 1978.

(iv) The training program shall include informing each employee of:

(a) The information contained in the substance information sheet for coke oven emissions (Appendix A);

(b) The purpose, proper use and limitations of respiratory protective devices required in accordance with paragraph **(g)** of this section;

(c) The purpose for and a description of the medical surveillance program required by paragraph **(j)** of this section including information on the occupational safety and health hazards associated with exposure to coke oven emissions;

(d) A review of all written procedures and schedules required under paragraph **(f)** of this section; and

(e) A review of this standard.

(2) Access to training materials.

(i) The employer shall make a copy of this standard and its appendixes readily available to all employees who are employed in the regulated area.

(ii) The employer shall provide, upon request, all materials relating to the employee information and training program to the Secretary and the Director.

Bloodborne pathogens 1910.1039(g)(2)(i) through (ix)

(2) Information and Training. **(i)** Employers shall ensure that all employees with occupational exposure participate in a training program which must be provided at no cost to the employee and during working hours.

(ii) Training shall be provided as follows:

(A) At the time of initial assignment to tasks where occupational exposure may take place;

(B) Within 90 days after the effective date of the standard; and

(C) At least annually thereafter.

(iii) For employees who have received training on bloodborne pathogens in the year preceding the effective date of the standard, only training with respect to the provisions of the standard which were not included need be provided.

(iv) Annual training for all employees shall be provided within one year of their previous training.

(v) Employers shall provide additional training when changes such as modification of tasks or procedures or institution of new tasks or procedures affect the employee's occupational exposure. The additional training may be limited to addressing the new exposures created.

(vi) Material appropriate in content and vocabulary to educational level, literacy and language of employees shall be used.

(vii) The training program shall contain at a minimum of the following elements:

(A) An accessible copy of the regulatory text of this standard and an explanation of its contents;

(B) A general explanation of the epidemiology and symptoms of bloodborne diseases;

(C) An explanation of the modes of transmission of bloodborne pathogens;

(D) An explanation of the employer's exposure control plan and the means by which the employee can obtain a copy of the written plan;

(E) An explanation of the appropriate methods for recognizing tasks and other activities that may involve exposure to blood and other potentially infectious material;

(F) An explanation of the use and limitations of methods that will prevent or reduce exposure including appropriate engineering controls, work practices and personal protective equipment;

(G) Information on the types, proper use, location, removal, handling, decontamination and disposal of personal protective equipment;

(H) An explanation of the basis of selection of personal protective equipment;

(I) Information on the Hepatitis B vaccine, including information on its efficacy, safety, method of administration, the benefits of being vaccinated and that the vaccine and vaccination will be offered free of charge;

(J) Information on the appropriate actions to take and persons to contact in an emergency involving blood or other potentially infectious materials;

(K) An explanation of the procedure to follow if an exposure incident occurs, including the method of reporting the incident and the medical follow up that will be made available;

(L) Information on the post exposure evaluation and follow up that the employer is required to provide for the employee following an exposure incident;

(M) An explanation of the signs and labels and/or color coding required by paragraph **(g)(1)**; and

(N) An opportunity for interactive questions and answers with the person conducting the training session.

(viii) The person conducting the training shall be knowledgeable in the subject matter covered by the elements contained in the training program as it relates to the workplace that the training will address.

(ix) Additional Initial Training for Employees in HIV and HBV Laboratories and Production Facilities. Employees in HIV or HBV research laboratories and HIV or HBV production facilities shall receive the following initial training in addition to the above training requirements.

(A) The employer shall assure that employees demonstrate proficiency in standard microbiological practices and techniques and in the practices and operations specific to the facility before being allowed to work with HIV and HBV.

(B) The employer shall assure that employees have prior experience in the handling of human pathogens or tissue cultures before working with HIV or HBV.

(C) The employer shall provide a training program to employees who have no prior experience in handling human pathogens. Initial work activities shall not include the handling of infectious agents. A progression of work activities shall be assigned as techniques are learned and proficiency is developed. The employer shall assure that employees participate in work activities involving infectious agents only after proficiency has been demonstrated.

Cotton Dust 1910.1043(i)(1) and (2)

(1) Training program.

(i) The employer shall provide a training program for all employees in all workplaces where cotton dust is present and shall assure that each employee in these workplaces is informed of the following:

(a) The specific nature of the operations which could result in exposure to cotton dust at or above the permissible exposure limit;

(b) The measures, including work practices required by paragraph (g) of this section, necessary to protect the employee from exposure in excess of the permissible exposure limit;

(c) The purpose, proper use and limitations of respirators required by paragraph (f) of this section;

(d) The purpose for and a description of the medical surveillance program required by paragraph **(h)** of this section and other information which will aid exposed employees in understanding the hazards of cotton dust exposures; and

(e) The contents of this standard and its appendixes.

(ii) The training program shall be provided prior to initial assignment and shall be repeated at least annually.

(2) Access to training materials.

(i) Each employer shall post a copy of this section with its appendixes in a public location at the workplace and shall, upon request, make copies available to employees.

(ii) The employer shall provide all materials relating to the employee training and information program to the Assistant Secretary and the Director upon request.

(iii) In addition to the information required by paragraph **(i)(1)**, the employer shall include as part of the training program, and shall distribute to employees, any materials pertaining to the Occupational Safety and Health Act, the regulations issued pursuant to that Act and this cotton dust standard which are made available to the employer by the Assistant Secretary.

1,2–Dibromo-3–Chloropropane 1910.1044(n)(1) and (2)

(1) Training program.

(i) The employer shall institute a training program for all employees who may be exposed to DBCP and shall assure their participation in such training program.

(ii) The employer shall assure that each employee is informed of the following:

(a) The information contained in Appendix A.

(b) The quantity, location, manner of use, release or storage of DBCP and the specific nature of operations which could result in exposure to DBCP as well as any necessary protective steps;

(c) The purpose, proper use and limitations of respirators;

(d) The purpose and description of the medical surveillance program required by paragraph **(m)** of this section; and

(e) A review of this standard including appendixes.

(2) Access to training materials.

(i) The employer shall make a copy of this standard and its appendixes readily available to all affected employees.

(ii) The employer shall provide, upon request, all materials relating to the employee information and training program to the Assistant Secretary and the Director.

Acrylonitrile (Vinyl Cyanide) 1910.1045(o)(1) and (2)

(1) Training program.

(i) By January 2, 1979 the employer shall institute a training program and assure the participation of all employees exposed to Acrylonitrile above the action level, all employees whose exposures are maintained below the action level by engineering and work practice controls, and all employees subject to potential skin or eye contact with liquid.

(ii) Training shall be provided at the time of initial assignment or upon institution of the training program and at least annually thereafter and the employer shall assure that each employee is informed of the following:

(A) The information contained in Appendixes A and B (Editor's Note: See Federal Register, Vol. 43, No. 192, Oct. 3, 1978, pp. 45813–45815);

(B) The quantity, location, manner of use, release or storage of AN and the specific nature of operations which could result in exposure to AN, as well as any necessary protective steps;

(C) The purpose, proper use and limitations of respirators and protective clothing;

(D) The purpose and a description of the medical surveillance program required by paragraph **(n)** of this section;

(E) The emergency procedures developed as required by paragraph **(i)** of this section;

(F) Engineering and work practice controls, their function and the employee's relationship to these controls; and

(G) A review of this standard.

(2) Access to training materials.

(i) The employer shall make a copy of this standard and its appendixes readily available to all affected employees.

(ii) The employer shall provide, upon request, all materials relating to the employee information and training program to the Assistant Secretary and the Director.

Ethylene Oxide 1910.1047(j)(3)(i–iii)

(3) Information and training. **(i)** The employer shall provide employees who are potentially exposed to EtO at or above the action level with information and training on EtO at the time of initial assignment and at least annually thereafter.

(ii) Employees shall be informed of the following:

(A) The requirements of this section with an explanation of its contents including Appendixes A and B;

(B) Any operations in their work area where EtO is present;

(C) The location and availability of the written EtO final rule; and

(D) The medical surveillance program required by paragraph **(i)** of this section with an explanation of the information in Appendix C.

(iii) Employer training shall include at least:

(A) Methods and observations that may be used to detect the presence or release of EtO in the work area (such as monitoring conducted by the employer, continuous monitoring devices, etc.);

(B) The physical and health hazards of EtO;

(C) The measures employees can take to protect themselves from hazards associated with EtO exposure including specific procedures the employer has implemented to protect employees from exposure to EtO, such as work practices, emergency procedures and personal protective equipment to be used; and

(D) The details of the hazard communication program developed by the employer including an explanation of the labeling system and how employees can obtain and use the appropriate hazard information.

Occupational Exposure to Hazardous Chemicals in Laboratories 1910.1450(f)(4)(i)(c)

(4) Training**. (i)** Employee training shall include:

(c) The measures employees can take to protect themselves from these hazards including specific procedures the employer has implemented to protect employees from exposure to hazardous chemicals, such as appropriate work practices, emergency procedures and personal protective equipment to be used.

A-5 CONSTRUCTION TRAINING REQUIREMENTS

A-5.1 General

The following training requirements have been excerpted from Title 29, Code of Federal Regulations Part 1926. Note that in addition to these requirements, Part 1910, relating to general industry, also contains applicable training standards.

General Safety and Health Provisions 1926.20(b)(2) and (4)

(2) Such programs [as may be necessary to comply with this part] shall provide for frequent and regular inspections of the job sites, materials and equipment to be made by competent persons [capable of identifying existing and predictable hazards in the surroundings or working conditions which are unsanitary, hazardous or dangerous to employees and who have authorization to take prompt corrective measures to eliminate them] designated by the employers.

(4) The employer shall permit only those employees qualified [one who, by possession of a recognized degree certificate or professional standing or who by extensive knowledge, training and experience, has successfully demonstrated his ability to solve or resolve problems relating to the subject matter, the work or the project] by training or experience to operate equipment and machinery.

Safety Training and Education 1926.21(a)

(a) General requirements. The Secretary shall, pursuant to section 107(f) of the Act, establish and supervise programs for the education and training of employers and employees in the recognition, avoidance and prevention of unsafe conditions in employments covered by the Act.

1926.21(b)(1) through (6)

(1) The employer should avail himself of the safety and health training programs the Secretary provides.

(2) The employer shall instruct each employee in the recognition and avoidance of unsafe conditions and the regulations applicable to his work environment to control or eliminate any hazards or other exposure to illness or injury.

(3) Employees required to handle or use poisons, caustics and other harmful substances shall be instructed regarding their safe handling and use and be made aware of the potential hazards, personal hygiene and personal protective measures required.

(4) In job site areas where harmful plants or animals are present, employees who may be exposed shall be instructed regarding the potential hazards and how to avoid injury and the first aid procedures to be used in the event of injury.

(5) Employees required to handle or use flammable liquids, gasses or toxic materials shall be instructed in the safe handling and use of these materials and made aware of the specific requirements contained in Subparts D, F and other applicable subparts of this part.

(6) (i) All employees required to enter into confined or enclosed spaces shall be instructed as to the nature of the hazards involved, the necessary precautions to be taken and in the use of protective and emergency equipment required. The employer shall comply with any specific regulations that apply to work in dangerous or potentially dangerous areas.

(6) (ii) For purposes of subdivision **(i)** of this subparagraph, "confined or enclosed space" means any space having a limited means of egress, which is subject to the accumulation of toxic or flammable contaminants or has an oxygen deficient atmosphere. Confined or enclosed spaces include, but are not limited to, storage tanks, process vessels, bins, boilers, ventilation or exhaust ducts, sewers, underground utility vaults, tunnels, pipelines and open top spaces more than 4 feet in depth such as pits, tubs, vaults and vessels.

Medical Services and First Aid 1926.50(c)

(c) In the absence of an infirmary, clinic, hospital or physician that is reasonably accessible in terms of time and distance to the worksite which is available for the treatment of injured employees, a person who has a valid certificate in first aid training from the U.S. Bureau of Mines, the American Red Cross or equivalent training that can be verified by documentary evidence, shall be available at the worksite to render first aid.

Ionizing Radiation 1926.53(b)

(b) Any activity which involves the use of radioactive materials or X-rays, whether or not under license from the Atomic Energy Commission [Nuclear Regulatory Commission] shall be performed by competent persons specifically trained in the proper and safe operation of such equipment. In the case of materials used under Commission license, only persons actually licensed or competent persons under the direction and supervision of the licensee shall perform such work.

Nonionizing Radiation 1926.54(a) and (b)

(a) Only qualified and trained employees shall be assigned to install, adjust and operate laser equipment.

(b) Proof of qualification of the laser equipment operator shall be available and in possession of the operator at all times.

Gases, Vapors, Fumes, Dusts and Mists 1926.55(b)

(b) To achieve compliance with paragraph (a) of this section, administrative or engineering controls must first be implemented whenever feasible. When such controls are not feasible to achieve full compliance, protective equipment or other protective measures shall be used to keep the exposure of employees to air contaminants within the limits prescribed in this section. Any equipment and technical measures used for this purpose must first be approved for each particular use by a competent industrial hygienist or other technically qualified person. Whenever respirators are used, their use shall comply with ss 1926.103.

Asbestos, Tremolite, Anthophyllite and Actinolite 1926.58(k)(3)(4)

(3) Employee information and training. **(i)** The employer shall institute a training program for all employees exposed to airborne concentrations of asbestos, tremolite, anthophyllite, actinolite or a combination of these minerals in excess of the action level and shall ensure their participation in the program.

(ii) Training shall be provided prior to or at the time of initial assignment, unless the employee has received equivalent training within the previous 12 months and at least annually thereafter.

(iii) The training program shall be conducted in a manner that the employee is able to understand. The employer shall ensure that each such employee is informed of the following:

(A) Methods of recognizing asbestos, tremolite, anthophyllite, actinolite exposure;

(B) The health effects associated with asbestos, tremolite, anthophyllite, actinolite exposure;

(C) The relationship between smoking and asbestos, tremolite, anthophyllite, actinolite in producing lung cancer;

(D) The nature of operations that could result in exposure to asbestos, tremolite, anthophyllite, actinolite, the importance of necessary protective controls to minimize exposure including, as applicable, engineering controls, work practices, respirators, housekeeping procedures, hygiene facilities, protective clothing, decontamination procedures, emergency procedures and waste disposal procedures and any necessary instruction in the use of these controls and procedures;

(E) The purpose, proper use, fitting instructions and limitations of respirators as required by 29 CFR 1910.134;

(F) The appropriate work practices for performing the asbestos, tremolite, anthophyllite or actinolite job; and

(G) Medical surveillance program requirements.

(H) A review of this standard, including appendixes.

(4) Access to training materials. **(i)** The employer shall make readily available to all affected employees without cost all written materials relating to the employee training program including a copy of this regulation.

(ii) The employer shall provide to the Assistant Secretary and the Director, upon request, all information and training materials relating to the employee information and training program.

Hearing Protection 1926.101(b)

(b) Ear protective devices inserted in the ear shall be fitted or determined individually by competent persons.

Respiratory Protection 1926.103(c)(1)

(1) Employees required to use respiratory protective equipment approved for use in atmospheres immediately dangerous to life shall be thoroughly trained in its use. Employees required to use other types of respiratory protective equipment shall be instructed in the use and limitations of such equipment.

Fire Protection 1926.150(a)(5)

(5) As warranted by the project, the employer shall provide a trained and equipped firefighting organization (Fire Brigade) to assure adequate protection to life. "Fire brigade" means an organized group of employees that are knowledgeable, trained and skilled in the safe evacuation of employees during emergency situations and in assisting in firefighting operations.

1926.150(c)(1)(viii)

(viii) Portable fire extinguishers shall be inspected periodically and maintained in accordance with Maintenance and Use of Portable Fire Extinguishers, NFPA No. 10A-1970.

From ANSI Standard 10A-19780. The owner or occupant of a property in which fire extinguishers are located has an obligation for the care and use of these extinguishers at all times. By doing so, he is contributing to the protection of life and property. The nameplate(s) and instruction manual should be read and thoroughly understood by all persons who may be expected to use extinguishers.

1120. To discharge this obligation he should give proper attention to the inspection, maintenance and recharging of this fire protective equipment. He should also train his personnel in the correct use of fire extinguishers on the different types of fires which may occur on his property.

3020. Persons responsible for performing maintenance operations come from three major groups:

Trained industrial safety or maintenance personnel.
Extinguisher service agencies.
Individual owners (e.g., self employed. . .).

Signaling 1926.201(a)(2)

(2) Signaling directions by flagmen shall conform to American National Standards Institute D6.1–1971, Manual on Uniform Traffic Control Devices for Streets and Highways.

Powder-Operated Hand Tools 1926.302(e)(1) and (12)

(1) Only employees who have been trained in the operation of the particular tool in use shall be allowed to operate a powder actuated tool.

(12) Powder-actuated tools used by employees shall meet all other applicable requirements of American National Standard Institute, A10.3–1970, Safety Requirements for Explosive Actuated Fastening Tools.

Woodworking Tools 1926.304(f)

(f) Other requirements. All woodworking tools and machinery shall meet other applicable requirements of American National Standards Institute, 01.1–1961,

Safety Code for Woodworking Machinery.

From ANSI Standard 01.1–1961, Selection and Training of Operators. Before a worker is permitted to operated any woodworking machine, he shall receive instructions in the hazards of the machine and the safe method of its operation. Refer to A9.7 of the Appendix.

"A9.7 Selection and Training of Operators. Operation of Machines, Tools and Equipment. General.

"**(1)** Learn the machine's applications and limitations, as well as the specific potential hazards peculiar to this machine. Follow available operating instructions and safety rules carefully.

"**(2)** Keep working area clean and be sure adequate lighting is available.

"**(3)** Do not wear loose clothing, gloves, bracelets, necklaces or ornaments. Wear face, eye, ear, respiratory and body protection devices, as indicated for the operation or environment.

"**(4)** Do not use cutting tools larger or heavier than the machine is designed to accommodate. Never operate a cutting tool at greater speed than recommended.

"**(5)** Keep hands well away from saw blades and other cutting tools. Use a push stock or push block to hold or guide the work when working close to cutting tool.

"**(6)** Whenever possible, use properly locked clamps, jig or vise to hold the work.

"**(7)** Combs (feather boards) shall be provided for use when an applicable guard cannot be used.

"**(8)** Never stand directly in line with a horizontally rotating cutting tool. This is particularly true when first starting a new tool or a new tool is initially installed on the arbor.

"**(9)** Be sure the power is disconnected from the machine before tools are serviced.

"**(10)** Never leave the machine with the power on.

"**(11)** Be positive that hold-downs and antikickback devices are positioned properly and that the workpiece is being fed through the cutting tool in the right direction.

"**(12)** Do not use a dull, gummy, bent or cracked cutting tool.

"**(13)** Be sure that keys and adjusting wrenches have been removed before turning power on.

"**(14)** Use only accessories designed for the machine.

"**(15)** Adjust the machine for minimum exposure of cutting tool necessary to perform the operation."

Gas Welding and Cutting 1926.350(d)(1) through (6)

(d) Use of fuel gas. The employer shall thoroughly instruct employees in the safe use of fuel gas as follows:

(1) Before a regulator to a cylinder valve is connected, the valve shall be opened slightly and closed immediately. (This action is generally termed "cracking" and is intended to clear the valve of dust or dirt that might otherwise enter the regulator.) The person cracking the valve shall stand to one side of the outlet, not in front of it.

The valve of a fuel gas cylinder shall not be cracked where the gas would reach welding work, sparks, flame or other possible sources of ignition.

(2) The cylinder valve shall always be opened slowly to prevent damage to the regulator. For quick closing, valves on fuel gas cylinders shall not be opened more than 1–1/2 turns. When a special wrench is required, it shall be left in position on the stem of the valve while the cylinder is in use so that the fuel gas flow can be shut off quickly in case of an emergency. In the case of manifold or coupled cylinders, at least one such wrench shall always be available for immediate use. Nothing shall be placed on top of a fuel gas cylinder when in use which may damage the safety device or interfere with the quick closing of the valve.

(3) Fuel gas shall not be used from cylinders through torches or other devices which are equipped with shutoff valves without reducing the pressure through a suitable regulator attached to the cylinder valve or manifold.

(4) Before a regulator is removed from a cylinder valve, the cylinder valve shall always be closed and the gas released from the regulator.

(5) If, when the valve on a fuel gas cylinder is opened, there is found to be a leak around the valve stem, the valve shall be closed and the gland nut tightened. If this action does not stop the leak, the use of the cylinder shall be discontinued, and it shall be properly tagged and removed from the work area. In the event that fuel gas should leak from the cylinder valve, rather than from the valve stem, and the gas cannot be shut off, the cylinder shall be properly tagged and removed from the work area. If a regulator attached to a cylinder valve will effectively stop a leak through the valve seat, the cylinder need not be removed from the work area.

(6) If a leak should develop at a fuse plug or other safety device, the cylinder shall be removed from the work area.

1926.350(j)

(j) Additional rules. For additional details not covered in this subpart, applicable technical portions of American National Standards Institute, Z49.1–1967, Safety in Welding and Cutting, shall apply.

From ANSI Standard Z49.1–1967, Fire Watch Duties. Fire watchers shall be trained in the use of fire extinguishing equipment. They shall be familiar with facilities for sounding an alarm in the event of a fire. They shall watch for fires in all exposed areas, try to extinguish them only when obviously within the capacity of the equipment available, or otherwise sound the alarm. A fire watch shall be maintained for at least a half hour after completion of welding or cutting operations to detect and extinguish possible smoldering fires.

Arc Welding and Cutting 1926.351(d)(1) through (5)

(d) Operating instructions. Employers shall instruct employees in the safe means of arc welding and cutting as follows:

(1) When electrode holders are to be left unattended, the electrodes shall be removed and the holders shall be so placed or protected that they cannot make electrical contact with employees or conducting objects.

(2) Hot electrode holders shall not be dipped in water; to do so may expose the arc welder or cutter to electric shock.

(3) When the arc welder or cutter has occasion to leave his work or to stop work for any appreciable length of time, or when the arc welding or cutting machine is to be moved, the power supply switch to the equipment shall be opened.

(4) Any faulty or defective equipment shall be reported to the supervisor.

(5) Other requirements as outlined in Article 630, National Electrical Code, NFPS 70–1971; ANSI C1–1971 (Rev. of 1968), Electric Welders, shall be used when applicable.

Fire Prevention 1926.352(e)

(e) When the welding, cutting or heating operation is such that normal fire prevention precautions are not sufficient, additional personnel shall be assigned to guard against fire while the actual welding, cutting or heating operation is being performed, and for a sufficient period of time after completion of the work to ensure that no possibility of fire exists. Such personnel shall be instructed as to the specific anticipated fire hazards and how the firefighting equipment provided is to be used.

Welding, Cutting and Heating in Way of Preservative Coatings 1926.354(a)

(a) Before welding, cutting or heating is commenced on any surface covered by a preservative coating whose flammability is not known, a test shall be made by a competent person to determine its flammability. Preservative coatings shall be considered to be highly flammable when scrapings burn with extreme rapidity.

Ground Fault Protection 1926.404(b)(iii)(B)

(iii)(B) The employer shall designate one or more competent persons (as defined in ss 1926.32(f)) to implement the program.

Scaffolding 1926.451(a)(3)

(3) No scaffold shall be erected, moved, dismantled or altered except under the supervision of competent persons.

1926.451(b)(16)

(16) All wood pole scaffolds 60 feet or less in height shall be constructed and erected in accordance with Tables L-4 to 10. If they are over 60 feet in height, they shall be designed by a qualified engineer competent in this field and it shall be constructed and erected in accordance with such design.

1926.451(c) (4) and (5)

(4) Tube and coupler scaffolds shall be limited in heights and working levels to those permitted in Tables L-10, 11 and 12. Drawings and specifications of all tube and coupler scaffolds above the limitations in Tables L10, 11 and 12 shall be designed by a qualified engineer competent in this field.

(5) All tube and coupler scaffolds shall be constructed and erected to support four times the maximum intended loads, as set forth in Tables L-10,11 and 12 or as set forth in the specifications by a licensed professional engineer competent in this field.

1926.451(d)(9)

(9) Drawings and specifications for all frame scaffolds over 125 feet in height above the base plates shall be designed by a registered professional engineer.

1926.451(g)(3)

(3) Unless outrigger scaffolds are designed by a registered engineer competent in this field, they shall be constructed and erected in accordance with Table L-13.

Outrigger scaffolds designed by a registered professional engineer shall be constructed and erected in accordance with such design.

1926.451(h)(6) and (14)

(h) Masons' adjustable multiple point suspension scaffolds.

(6) Where the overhang exceeds 6 feet 6 inches, outrigger beams shall be composed of stronger beams or multiple beams and be installed under the supervision of a competent person.

(14) Each scaffold shall be installed or relocated under the supervision of a competent person.

1926.451(k)(10)

(k) Single point adjustable suspension scaffolds.

(10) For additional details not covered in this paragraph, applicable technical portions of American National Standards Institute, A120.1–1970, Power Operated Devices for Exterior Building Maintenance Powered Platforms, shall be used.

From ANSI Standard A120.1–1970. Qualified Operators. Powered platform shall be operated only by qualified persons who have been instructed in the operation and in the inspection with respect to safe operating condition of the particular powered platform to be operated.

Guarding of Low Pitched Roof Perimeters During the Performance of Built Up Roofing Work 1926.500(g)(6)

(6) Training. **(i)** The employer shall provide a training program for all employees engaged in built up roofing work so that they are able to recognize and deal with the hazards of falling associated with working near a roof perimeter. The employees shall also be trained in the safety procedures to be following in order to prevent such falls.

(ii) The employer shall assure that employees engaged in built up roofing work have been trained and instructed in the following areas:

(a) The nature of fall hazards in the work area near a roof edge;

(b) The function, use and operation of the MSS system, warning line and safety monitoring systems to be used to be used;

(c) The correct procedures for erecting, maintaining, and disassembling the systems to be used;

(d) The role of each employee in the safety monitoring system when this system is used;

(e) The limitations of the use of mechanical equipment; and

(f) The correct procedures for the handling and storage of equipment and materials.

(iii) Training shall be provided for each newly hired employee, and for all other employees as necessary, to assure that employees maintain proficiency in the areas listed in paragraph **(g)(6)(ii)** of this section.

Cranes and Derricks 1926.550(a)(1), and (5)

(1) The employer shall comply with the manufacturer's specifications and limitations applicable to the operation of any and all cranes and derricks. Where manufacturer's specifications are not available, the limitations assigned to the equipment shall be based on the determinations of a qualified engineer competent in this field and such determinations will be appropriately documented and recorded. Attachments used with cranes shall not exceed the capacity, rating or scope recommended by the manufacturer.

(5) The employer shall designate a competent person who shall inspect all machinery and equipment prior to each use, and during use, to make sure it is in safe operating condition. Any deficiencies shall be repaired, or defective parts replaced, before continued use.

1926.550(g)(4)

(g)(4) Personnel Platforms. The personnel platform and suspension system shall be designed by a qualified engineer or a qualified person competent in structural design.

1926.550(g)(5)(iv)

(g)(iv) A visual inspection of the crane or derrick, rigging, personnel platform and the crane or derrick base support or ground shall be conducted by a competent person immediately after the trial lift to determine whether the testing has exposed any defect or produced any adverse effect upon any component or structure.

Material Hoists, Personnel Hoists and Elevators 1926.552(a)(1)

(1) The employer shall comply with the manufacturer's specifications and limitations applicable to the operation of all hoists and elevators. Where manufacturer's specifications are not available, the limitations assigned to the equipment shall be based on the determinations of a professional engineer competent in the field.

1926.552(b)(7)

(7) All material hoist towers shall be designed by a licensed professional engineer.

1926.552(c)(15) and (17) (i)

(c) Personnel hoists.

(15) Following assembly and erection of hoists, and before being put in service, an inspection and test of all functions and safety devices shall be made under the supervision of a competent person. A similar inspection and test is required following major alteration of an existing installation. All hoists shall be inspected and tested at not more than 3 month intervals. Records shall be maintained and kept on file for the duration of the job.

(17)(i) Personnel hoists used in bridge tower construction shall be approved by a registered professional engineer and erected under the supervision of a qualified engineer competent in this field.

Material Handling Equipment 1926.602(c)(1)(vi)

(c) Lifting and hauling equipment (other than equipment covered under Subpart N of this part).

(1)(vi) All industrial trucks in use shall meet the applicable requirements of design, construction, stability, inspection, testing, maintenance and operation, as defined in American National Standards Institute B56.1–1969, Safety Standards for Powered Industrial Trucks.

From ANSI Standard B56.1–1969. Operator Training. Only trained and authorized operators shall be permitted to operate a powered industrial truck. Methods shall be devised to train operators in the safe operation of powered industrial trucks. Badges or other visual indication of the operator's authorization should be displayed at all times during work period.

Site Clearing 1926.604(a)(1)

(1) Employees engaged in site clearing shall be protected from hazards of irritant and toxic plants and suitably instructed in the first aid treatment available.

Excavations General Protection Requirements (Excavations, Trenching and Shoring) 1926.650(i)

(i) Daily inspections of excavations shall be made by a competent person. If evidence of possible cave ins or slides is apparent, all work in the excavation shall cease until the necessary precautions have been taken to safeguard the employees.

1926.651(i)(2)

(iii) A registered professional engineer has approved the determination that the structure is sufficiently removed from the excavation so as to be unaffected by the excavation activity; or

(iv) A registered professional engineer has approved the determination that such excavation work will not pose a hazard to employees.

(k) Support systems shall be planned and designed by a qualified person when excavation is in excess of 20 feet in depth, adjacent to structures or improvements or subject to vibration or ground water.

(o) If the stability of adjoining buildings or walls is endangered by excavations, shoring, bracing or underpinning shall be provided as necessary to insure their safety. Such shoring, bracing or underpinning shall be inspected daily or more often, as conditions warrant, by a competent person and the protection effectively maintained.

(x) Where ramps are used for employees or equipment, they shall be designed and constructed by qualified persons in accordance with accepted engineering requirements.

Concrete and Masonry Construction 1926.701(a)

(a) No construction loads shall be placed on a concrete structure or a portion of a concrete structure unless the employer determines, based on information received from a person who is qualified in structural design, that the structure or portion of the structure is capable of supporting the loads.

"Shoring" 1926.703(b)(8)(i)

(i) The design of the shoring shall be prepared by a qualified designer and the erected shoring shall be inspected by an engineer qualified in structural design.

Bolting, Riveting, Fitting-Up and Plumbing-Up 1926.752(d)(4)

(4) Plumbing-up guys shall be removed only under the supervision of a competent person.

Underground Construction 1926.800(d)

(d) Safety instruction. All employees shall be instructed in the recognition and avoidance of hazards associated with underground construction activities including, where appropriate, the following subjects:

(1) Air monitoring;
(2) Ventilation;
(3) Illumination;
(4) Communications;
(5) Flood Control;
(6) Mechanical equipment;
(7) Personal protective equipment;
(8) Fire prevention and protection; and
(9) Emergency procedures including evacuation plans and check-in/check-out systems.

Compressed Air 1926.803(a)(1) and (2)

(1) There shall be present, at all times, at least one competent person designated by and representing the employer, who shall be familiar with this subpart in all respects and responsible for full compliance with these and other applicable subparts.

(2) Every employee shall be instructed in the rules and regulations which concern his safety or the safety of others.

1926.803(b)(1) and (10)(xii)

(1) There shall be retained one or more licensed physicians familiar with and experienced in the physical requirements and the medical aspects of compressed air work and the treatment of decompression illness. He shall be available at all times while work is in progress in order to provide medical supervision of employees employed in compressed air work. He shall himself by physically qualified and be willing to enter a pressurized environment.

(10) The medical lock shall: **(xii)** Be in constant charge of an attendant under the direct control of the retained physician. The attendant shall be trained in the use of the lock and suitably instructed regarding steps to be taken in the treatment of employees exhibiting symptoms compatible with a diagnosis of decompression illness.

1926.803(e)(1)

(1) Every employee going under air pressure for the first time shall be instructed on how to avoid excessive discomfort.

1926.803(h)(1)

(1) At all times there shall be a thoroughly experienced, competent and reliable person on duty at the air control valves as a gauge tender who shall regulate the

pressure in the working areas. During tunneling operations, one gauge tender may regulate the pressure in not more than two headings provided that the gauge and controls are all in one location. In caisson work, there shall be a gauge tender for each caisson.

Preparatory Operations 1926.850(a)

(a) Prior to permitting employees to start demolition operations, an engineering survey shall be made, by a competent person, of the structure to determine the condition of the framing, floors and walls and the possibility of unplanned collapse of any portion of the structure. Any adjacent structure where employees may be exposed shall also be similarly checked. The employer shall have in writing evidence that such a survey has been performed.

Chutes 1926.852(c)

(c) A substantial gate shall be installed in each chute at or near the discharge end. A competent employee shall be assigned to control the operation of the gate and the backing and loading of trucks.

Mechanical Demolition 1926.859(g)

(g) During demolition, continuing inspections by a competent person shall be made as the work progresses to detect hazards resulting from weakened or deteriorated floors or walls or loosened material. No employee shall be permitted to work where such hazards exist until they are corrected by shoring, bracing or other effective means.

General Provisions (Blasting and Use of Explosives) 1926.900(a)

(a) The employer shall permit only authorized and qualified persons to handle and use explosives.

1926.900(k)(3)(i)

(i) The prominent display of adequate signs warning against the use of mobile radio transmitters on all roads within 1,000 feet of blasting operations. Whenever adherence to the 1,000 foot distance would create an operational handicap, a competent person shall be consulted to evaluate the particular situation and alternative provisions may be made which are adequately designed to prevent any premature firing of electric blasting caps. A description of any such alternatives shall be reduced to writing and shall be certified as meeting the purposes of this subdivision by the competent person consulted. The description shall be maintained at the construction site during the duration of the work and shall be available for inspection by representatives of the Secretary of Labor.

1926.900(q)

(q) All loading and firing shall be directed and supervised by competent persons thoroughly experienced in this field.

Blaster Qualifications 1926.901(c), (d), and (e)

(c) A blaster shall be qualified by reason of training, knowledge or experience in the field of transporting, storing, handling and the use of explosives and have a working knowledge of state and local laws and regulations which pertain to explosives.

(d) Blasters shall be required to furnish satisfactory evidence of competency in handling explosives and performing in a safe manner the type of blasting that will be required.

(e) The blaster shall be knowledgeable and competent in the use of each type of blasting method used.

Surface Transportation of Explosives 1926.902(b) and (i)

(b) Motor vehicles or conveyances transporting explosives shall only be driven by, and be in the charge of, a licensed driver who is physically fit. He shall be familiar with the local, state and federal regulation governing the transportation of explosives.

(i) Each vehicle used for transportation of explosives shall be equipped with a fully charged fire extinguisher, in good condition. An Underwriters Laboratory approved extinguisher of not less than 10–ABC rating will meet the minimum requirement. The driver shall be trained in the use of the extinguisher on his vehicle.

Firing the Blast 1926.909(a)

(a) A code of blasting signals equivalent to Table U-1 shall be posted on one or more conspicuous places at the operation and all employees shall be required to familiarize themselves with the code and conform to it. Danger signs shall be placed at suitable locations.

Table U-1 Warning Signal - A 1–minute series of long blasts 5 minutes prior to blast signal.

Blast Signal - A series of short blasts 1 minute prior to the shot.

All-Clear Signal - A prolonged blast following the inspection of blast area.

General Requirements (Power Transmission and Distribution) 1926.950(d)(1)(ii), (vi) and (vii)

(1) When deenergizing lines and equipment operated in excess of 600 volts and the means of disconnecting from electric energy is not visibly open or visibly locked, the provisions of subdivisions **(i)** through **(vii)** of this subparagraph shall be complied with:

(ii) Notification and assurance from the designated employee [a qualified person delegated to perform specific duties under the conditions existing] shall be obtained that:

(a) All switches and disconnectors through which electric energy may be supplied to the particular section of line or equipment to be worked have been deenergized;

(b) All switches and disconnectors are plainly tagged indicating that men are at work;

(c) And that where design of such switches and disconnectors permits, they have been rendered inoperable.

(vi) When more than one independent crew requires the same line or equipment to be deenergized, a prominent tag for each such independent crew shall be placed on the line or equipment by the designated employee in charge.

(vii) Upon completion of work on deenergized lines or equipment, each designated employee in charge shall determine that all employees in his crew are clear, that protective grounds installed by his crew have been removed, and he shall report to the designated authority that all tags protecting his crew may be removed.

1926.950(d)(2)(ii)

(2) When a crew working on a line or equipment can clearly see that the means of disconnecting from electric energy are visibly open or visibly locked-out, the provisions of subdivisions **(i)** and **(ii)** of this paragraph shall apply:

(ii) Upon completion of work on deenergized lines or equipment, each designated employee in charge shall determine that all employees in his crew are clear, that protective grounds installed by his crew have been removed and he shall report to the designated authority that all tags protecting his crew may be removed.

1926.950(e)(1) and (2)

(1) The employer shall provide training or require that his employees are knowledgeable and proficient in:

(i) Procedures involving emergency situations, and

(ii) First aid fundamentals including resuscitation.

(2) In lieu of subparagraph (1) of this paragraph, the employer may comply with the provisions of ss 1926.950(c) regarding first aid requirements.

Overhead Cranes 1926.955(b)(8) and (d)(1)

(8) A designated employee shall be utilized to determine that required clearance is maintained in moving equipment under or near energized lines.

(1) Prior to stringing parallel to an existing energized transmission line, a competent determination shall be made to ascertain whether dangerous induced voltage buildups will occur, particularly during switching and ground fault conditions. When there is a possibility that such dangerous induced voltage may exist the employer shall comply with the provisions of subparagraphs **(2)** through **(9)** of this paragraph in addition to the provisions of paragraph **(c)** of this ss1926.955, unless the line is worked as energized.

1926.955(e)(1) and (4)

(1) Employees shall be instructed and trained in the live-line bare-hand technique and the safety requirements pertinent thereto before being permitted to use the technique on energized circuits.

(4) All work shall be personally supervised by a person trained and qualified to perform live-line bare-hand work.

Underground Lines 1926.956(b)(1)

(1) While work is being performed in manholes, an employee shall be available in the immediate vicinity to render emergency assistance as may be required. This shall not preclude the employee in the immediate vicinity from occasionally entering a manhole to provide assistance other than emergency. This requirement does not preclude a qualified employee [a person who by reason of experience or training is familiar with the operation to be performed and the hazards involved], working alone, from entering for brief periods of time, a manhole where energized cables or equipment are in service, for the purpose of inspection, housekeeping, taking readings or similar work is such work can be performed safely.

Construction in Energized Substations 1926.957(a)(1)

(1) When construction work is performed in an energized substation, authorization shall be obtained from the designated authorized person [a qualified person delegated to perform specific duties under the conditions existing] before work is started.

1926.957(d)(1)

(1) Work on or adjacent to energized control panels shall be performed by designated employees.

1926.957(e)(1)

(1) Use of vehicles, gin poles, cranes and other equipment in restricted or hazardous areas shall at all times be controlled by designated employees.

Ladders 1926.1053(b)(15)

(15) Ladders shall be inspected by a competent person for visible defects on a periodic basis and after any occurrence that could affect their safe use.

Training Provisions 1926.1060

The following training provisions clarify the requirements of ss 1926.21(b)(2), regarding the hazards addressed in subpart X.

(a) The employer shall provide a training program for each employee using ladders and stairways, as necessary. The program shall enable each employee to recognize hazards related to ladders and stairways and shall train each employee in the procedures to be followed to minimize these hazards.

(1) The employer shall ensure that each employee has been trained by a competent person in the following areas, as applicable:

(i) The nature of fall hazards in the work area;

(ii) The correct procedures for erecting, maintaining and disassembling the fall protection systems to be used;

(iii) The proper construction, use, placement and care in handling of all stairways and ladders;

(iv) The maximum intended load-carrying capacities of ladders used; and

(v) The standards contained in this subpart.

(b) Retraining shall be provided for each employee as necessary so that the employee maintains the understanding and knowledge acquired through compliance with this section.

Index

D

E

F

I

J

L

M

M.E. Sharpe, Inc. Software License Agreement

The Construction Safety and Loss Control Program Manual, by Andrew Civitello, Jr.

LICENSE AGREEMENT

This is a legal agreement between you, the end user, and M.E. Sharpe, Inc.

By breaking the seal on the package containing the *Quick*➜*Disk*, you have agreed to be bound by the terms of this agreement. If you do not agree to the terms of this agreement, do not open the package containing the diskette and return the diskette and its original packaging with the book to M.E. Sharpe, Inc.

If you open the seal of the *Quick*➜*Disk* you may not return the disk, unless for replacement, nor may you return the print edition of the Construction Safety and Loss Control Program Manual.

See Limited Warranty Section below regarding defective material and media.

M.E. Sharpe Inc. retains the ownership of this copy of software. This copy is licensed to you for use under the following conditions:

Permitted Uses/You May:

- Use the software on any compatible computer, provided the software is used on only one computer and by one user at a time.
- Use the software on a network, file server or virtual disk, provided that you have a copy of the documentation and an authorization device, and the program disk. You must employ the authorization device when using the software at your computer.

Prohibited Uses/You May Not:

- Make copies of the program disk.
- Distribute, rent, sub-license or lease the software or documentation.
- Alter, modify or adapt the software or documentation, including but not limited to, translating, decompiling, disassembling, or creating derivative works.

This license and your right to use the software automatically terminate if you fail to comply with any provision of this License Agreement.

Limited Warranty

M.E. Sharpe Inc. warrants that the program will substantially conform to the published specifications and to the documentation, provided that it is used on the computer hardware and with the operating system for which it was designed. M.E. Sharpe Inc. also warrants that the magnetic media on which the program is distributed and the documentation are free from defects in materials and workmanship. M.E. Sharpe Inc. will replace defective media at no charge, provided you return the item with the original packaging with the book to M.E. Sharpe Inc. within 90 days of the date of delivery.

Except as specifically provided above, M.E. Sharpe Inc. makes no warranty or representation, either express or implied, with respect to this program or documentation.

In no event will M.E. Sharpe Inc. be liable for direct, indirect, special, incidental, or consequential damages arising out of the use of or inability to use the program or documentation, even if advised of the possibility of such damages. Specifically, M.E. Sharpe is not responsible for any costs including, but not limited to, those incurred as result of lost profits or revenue, loss of use of the computer program, loss of data, the costs of recovering such programs or data, the cost of any substitute program, claims by third parties, or for other similar costs. In no case shall M.E. Sharpe's liability exceed the amount of the license fee (ie: cost of product).

The warranty and remedies set forth above are exclusive and in lieu of all others, oral or written, express or implied. No M.E. Sharpe Inc. dealer, distributor, agent, or employee is authorized to make any modification or addition to this warranty.

Some states do not allow the exclusion or limitation of implied warranties or limitation of liability for incidental or consequential damages, so the above limitation or exclusion may not apply to you. In the U.K. this warranty does not affect the statutory rights of a consumer.

General
M.E. Sharpe Inc. retains all rights not expressly granted. Nothing in this License Agreement constitutes a waiver of M.E. Sharpe's rights under the copyright laws or any other Federal or State law.

Should you have any questions concerning this License Agreement, you may contact M.E. Sharpe by writing to M.E. Sharpe, Inc., 80 Business Park Drive, Armonk, New York, 10504.

THE M.E. SHARPE SOFTWARE LICENSE AGREEMENT, DISCLAIMER, LIMITED WARRANTY AND INSTRUCTIONS ON HOW TO USE THE DISK, are also provided on the "Read Me" file on the enclosed disk.

How to use the Construction Safety and Loss Control Program Manual *Quick–Disk* :

Users of the ***Construction Safety and Loss Control Manual Quick—Disk*** are encouraged to use it in preparing their own company safety and loss control manuals for duplication and distribution within and only within your company.

The contents of the ***Construction Safety and Loss Control Manual*** *Quick–Disk* have been provided in ASCII text format which will allow any user to utilize the data with their preferred word processing software.

The *Quick–Disk* files can be opened in "text format" and will have a .TXT file extension. In order to open the *Quick–Disk* files, your word processing software must list all types of files (*.*) or files in text format.

[ ] Check the section HOW TO USE THIS MANUAL which appears in the Introduction of the book and follow the 10 Steps listed under the section heading "Procedure for Manual Development", and
[ ] Check the Section Use material which appears at thebeginning of each Sample Section throughout the book.

Although ASCII text is highly versatile, all text in the ***Construction Safety and Loss Control Program Manual*** *Quick–Disk* is provided in normal mode. No provisions have been made for bold or italic text. If you would like your manual to follow the style of the published text, you must apply boldface and italic type to your documents.

Tabs have been inserted in the ASCII format in the following areas:

As a paragraph indent;
As indents in a bulleted list;
As indents in numbered lists and outlines.

Most software programs will allow you to reformat the tabs according to your specifications.

Some of the chapter sections have pertinent notes, entitled "Note to Manual Preparer." (These "Notes" are in boldface italic type in the published text.) Each "Note" has been distinguished from the ***Construction Safety and Loss Control Program Manual*** text on this disk by the insertion of a line above and below the "Note." These "Notes" can be deleted from the disk once you are prepared to print the final copy.

The following chapters include special forms, which are not included on this disk: 2, 4, 6, 8, 10, 11, 14. These forms can be photocopied from the book and inserted in your manual. We have included a note in the computer text to alert the user where to insert each form, such as the following example: <INSERT A COPY OF THE FOLLOWING FORM FROM PAGE 41: (COMPANY) Jobsite Safety Inspection Report Form>.

Some elements of the disk, such as the tables, need special formatting. These elements have been flagged with notes, such as <SEE PAGE 239 FOR TABLE FORMAT>, to refer the user to the published manual.

Other elements of the disk need to be filled in by the user (such as your company's name). These items will become apparent upon careful reading of the ***Construction Safety and Loss Control Program Manual.***

In view of the appendix's total byte count, we have divided it into two separate files.

Remember to save your data frequently.